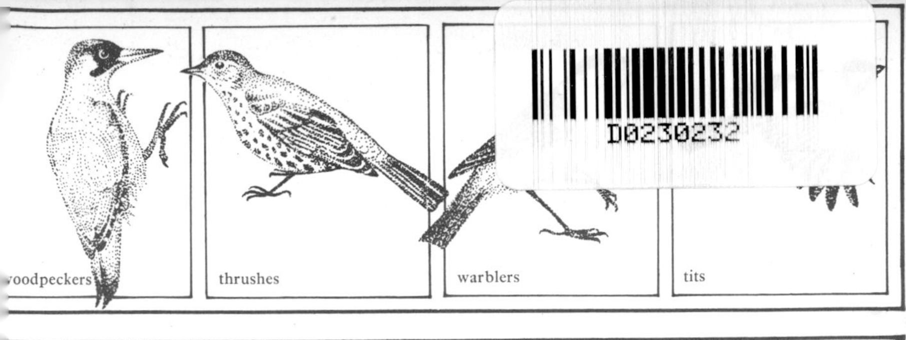

woodpeckers

thrushes

warblers

tits

10

insectivores

rodents

bats

carnivores

lagomorphs

ungulates

11

fungi

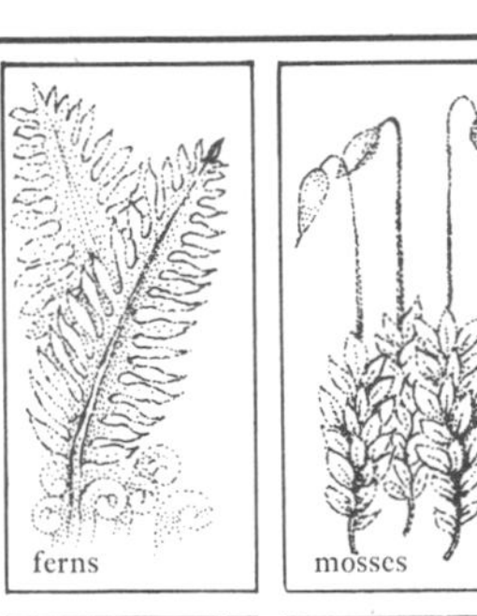

ferns

mosses

Ranunculaceae

Papaveraceae

Rosaceae

Leguminosae

Umbelliferae

Primulaceae

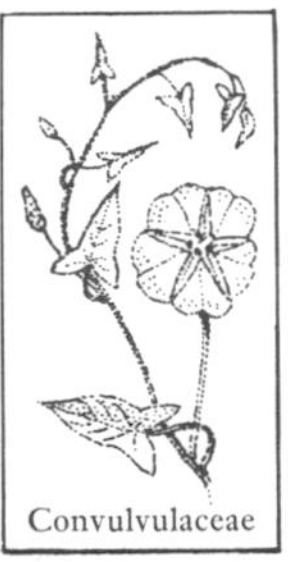

Convulvulaceae

Dipsacaceae

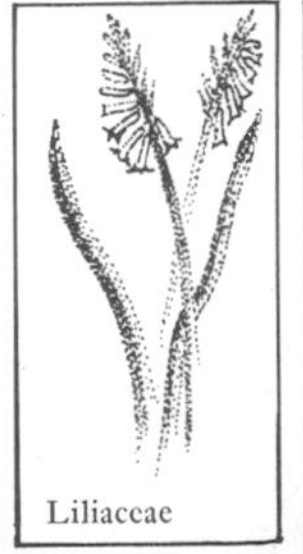

Liliaceae

THE NATURAL HISTORY OF BRITAIN
AND NORTHERN EUROPE

FIELDS AND LOWLANDS

DERRICK BOATMAN

Editors JAMES FERGUSON-LEES & BRUCE CAMPBELL

Contributors Franklyn and Margaret Perring (Plants);
Paul Whalley, Frederick Wanless, Edward Easton, Roger Lincoln,
Peter Mordan (Invertebrates); Tim Halliday (Amphibians and Reptiles);
James Ferguson-Lees (Birds); Gordon Corbet (Mammals)

Illustrators Deborah King (Plants); Joyce Tuhill (Invertebrates);
Hilary Burn (Amphibians, Reptiles, Birds and Mammals)

Hodder & Stoughton
LONDON · SYDNEY · AUCKLAND · TORONTO

This book was designed and produced by
George Rainbird Limited,
36 Park Street, London W1Y 4DE
for Hodder & Stoughton Limited,
Mill Road, Dunton Green,
Sevenoaks, Kent

House Editors: Karen Goldie-Morrison,
Linda Gamlin
Designer: Patrick Yapp
Indexer: Diana Blamire
Picture Researcher: Karen Gunnell
Cartographer: Tom Stalker Miller
Cover Illustrator: Hilary Burn
Endpapers Illustrator: Terry Callcut
Production: Bridget Walsh, Jane Collins

Printed and bound by
W. S. Cowell Limited,
28 Percy Street, London W1P 9FF

ISBN 0 340 23153 X

CONTENTS

FOREWORD

With increased travel and an expanding interest in Europe as a whole, many books and field guides on its natural history have been published in the last two decades, but most either treat a wide field in general terms or cover a single class or group of animals or plants. At the same time, inspired by the need for conservation, the pendulum is swinging back from the specialization of the post-war years to a wish for a fuller appreciation of all aspects of natural history. As yet, the traveller-naturalist has to be armed with a variety of volumes and, even then, has no means of understanding the interrelations of plants and animals. We believe that this new series will help to fill that gap.

The five books cover the whole of the northern half of Europe west of Russia and the Baltic States, and include Iceland: the limits are shown by the map on pages 68–69, which marks the individual countries, and the various sub-regions with the abbreviations used for them. Four of the volumes deal with (a) towns and gardens; (b) coniferous forests, heaths, moors, mountains and tundras; (c) lakes, rivers and freshwater marshes; and (d) coasts, dunes, sea-cliffs, saltmarshes, estuaries and the sea itself. This book is about broad-leaved woods, hedgerows, farmland, lowland grassland and downs. Thus, in broad groupings of related habitats, the series covers the whole rural and urban scene.

Each book is divided into two. The first half is an ecological essay about the habitats, with examples of plant and animal communities as illustrations of interrelationships. The second is a field guide of selected species, each illustrated and described, with its habitat, the part it plays in food webs, and its distribution. Obviously there are limitations: about 600 species are illustrated in each book, or around 3000 in the series, but the north European total is probably at least 50,000. Whereas good proportions of the characteristic vertebrates (148 mammals, 364 birds, 18 reptiles, 22 amphibians and more than 300 fishes) are included, some single *families* of insects have more species than the total of these; there are over 4000 different beetles in Britain alone, and probably 8–9000 in the whole of our area, while some 3500 plants are also native or naturalized in north Europe. On the other hand, the identification of many insects and some groups of plants is a matter for the specialist and we believe that many readers will be satisfied if they can identify these at the family level. So our list of invertebrates and plants is selective, but we hope that it will form a useful groundwork from which interest in particular groups can be developed.

All plants and animals are grouped into classes (*eg* Angiospermae, Insecta, Aves), orders (*eg* Campanulatae, Lepidoptera, Passeriformes), families (*eg*

Compositae, Nymphalidae, Turdidae) and genera (*eg Aster*, *Inachis*, *Turdus*), groups of increasingly close affinity. Each plant or animal has two scientific names, the first of which is the genus and the second the species. These are often considered to be outside the scope of a work of this kind, but many invertebrates and some plants have no vernaculars and, at the same time, such names are invaluable in showing relationships. Consequently, each species is given its scientific name at the first mention in each essay chapter and again in the field guide, where the family name is also inserted in capitals.

The specially commissioned colour paintings which illustrate the field guide are a delight in themselves. It has become customary to illustrate plants and animals in field guides as individual specimens, but here they are arranged in attractive compositions. Scale has had to suffer, but the sizes are always given in the facing descriptions, as are the correct food plants of the invertebrates.

The book on towns and gardens was concerned with a habitat virtually created by man in northern Europe over the past two thousand years. This book covers the original habitats most markedly changed by him. He has felled almost all the primitive broadleaved forest, replacing it often with the same species but organizing the woodland for his own purposes, as in the coppice-with-standards system once so prevalent in southern England. These modifications have until recently often been to the benefit of wildlife. Coppice-with-standards, allowing more light to reach the woodland floor, enriched it with flowering plants. Hedgerows, attenuated strips of woodland with an abundance of 'edge', encouraged a variety of sun-loving invertebrate animals as well as bringing forest birds into the open countryside and creating protected alleys through it. Grazing of lime-rich downlands produced a floriferous sward together with the food plants of butterflies and other insects. Cultivations spread a host of 'weeds', many of them alien to the region. Modern agricultural technology is fast changing all this, but enough of it remains to attract the naturalist and the visitor with an enquiring mind.

The author of the ecological essay, Dr Derrick Boatman, lectured for four years at Trinity College, Dublin, before coming to the University of Hull, where he is Reader in Ecology in the Department of Plant Biology. He has travelled widely in the region covered by this book, especially in France, the Netherlands, Germany, Sweden and Finland. One of his present research projects is the study of hedgerows in the former East Riding of Yorkshire, an area now so intensively farmed that it is impossible to tell what the natural vegetation was like; but from data he and his students are collecting it may be possible to work out the composition of the semi-natural vegetation in the Middle Ages. His background and present interests therefore make him a particularly suitable author for this book.

JAMES FERGUSON-LEES
BRUCE CAMPBELL

INTRODUCTION

This book concerns those parts of Europe where broadleaved forest is capable of forming the dominant feature of the landscape. The southern part of the area, stretching from the River Meuse in eastern France across central Germany to the River Oder in western Poland, consists of a series of ancient mountain ranges, now greatly denuded, known as the Central Uplands. Beyond the Oder, and continuing eastwards across southern Poland, are the Carpathians. Between the Uplands and the North Sea and Baltic coasts is a vast wedge-shaped area rarely higher than 200 m, which is known as the Northern Lowlands. At their western extremity, between the Ardennes and the coast of Belgium, these Lowlands are about 200 km wide, but along the eastern frontier of Poland their width is about 700 km. Most of the British Isles fall within the broadleaved forest zone, but Norway, Finland and all but the southern part of Sweden are occupied by coniferous forest.

Large-scale forest clearances

Until about 3000 BC, almost the whole of the Lowlands was covered by forest: broadleaved, mixed or, in some places, almost pure coniferous. By the time the Romans reached Britain, much of the higher, well-drained chalk and limestone land had apparently been cleared, but in the valleys forest was still extensive (though probably modified from its primeval state). It is estimated that, at the beginning of the Christian era, at least two-thirds of west-central Europe outside the mountains was still either forested or marshy.

Clearance of the lowland forests of western Europe seems to have begun between AD 500 and 800. By this time, heavy ploughs were being drawn by teams of oxen and these were very effective at turning the heavy, lowland soil. A rural pattern of large, open fields attached to hamlets took the place of the forests. Each peasant farmer had strips of land in each of the fields, as well as the right to graze cattle on the rough grassland attached to the village and to graze and take wood from any scrub or woodland that remained. Large areas of forest existed in Norman Britain, but many of these, used for hunting, probably occupied land which had previously been ploughed and were not therefore in the primeval state.

By the end of the Middle Ages, it is likely that all of the remaining areas of woodland in Britain south of the Scottish Highlands had come under some form of human management and were therefore markedly different from the

Lowland scene with arable land on the valley bottom, grassland on the poorer soils and scrub invading the steeper slopes. The hedgerows are no longer stock-proof; Hampshire.

natural state. The practice of coppicing shrubs had begun and domestic animals were allowed to graze and browse in the woods. By this time, however, a developing demand for large timber for ships and for charcoal used in the smelting of ores resulted in further devastation of the forest. About the middle of the 17th century, destruction of the Highland forest began and soon Britain had become the least afforested country in Europe, with a cover of less than five per cent.

In eastern Europe, destruction of the forest apparently started later and proceeded more slowly. For instance, the greater part of Poland was still under forest in the 11th century and even today the cover is about 20 per cent; of this, however, only about 12 per cent is broadleaved.

Natural regeneration in forests

In a mature forest, the amount of light reaching the floor may be so small that tree seedlings are unable to establish themselves. As an overmature tree begins to die back, however, the amount of illumination around it gradually increases and seedlings can grow.

Sometimes, sudden gaps appear in the canopy when, for instance, trees are blown down. On such occasions, the manner in which forest is re-established depends on a number of factors, notably the size of the gap and the composition of the surrounding vegetation.

When a species with an efficient means of seed dispersal is present near a gap, its seeds are likely to be distributed quickly over the whole area, with the result that its saplings become the only trees present. Thus, in a forest of beech *Fagus sylvatica* and ash *Fraxinus excelsior*, the latter's winged seeds may colonize gaps very quickly; ash trees also grow fast and, for a period, will become dominant. The degree of shade beneath an ash canopy is, however, less intense than beneath beech and consequently beech seedlings will be able gradually to invade the gap, become established among the ash trees and, because they live longer, eventually replace them. Thus re-establishment of forest often occurs by successive stages.

In the majority of forest types, the process of recolonization will be more complicated than this because of the presence of various herbs, scrambling and climbing plants, and shrubs. These will respond to the increased light resulting from, for example, the wind-throw of trees, and their growth may be so vigorous that recolonization by trees is delayed. Recolonization will occur eventually, however, and as the tree canopy develops, the vigour of the associated species will gradually be reduced. A walk through a piece of woodland may, in effect, take one backwards and forwards many times through various stages of a succession.

Nutrient cycles

More than 80 per cent of the weight of living, non-woody plant tissue is water. Of the remainder, more than half is carbon, absorbed from the atmosphere as a gas, carbon dioxide. Relatively small amounts of nitrogen, potassium, phosphorus, calcium and magnesium are present, together with minute quantities of trace elements, such as iron, copper, manganese and sulphur. All these are obtained from the soil, and in a natural ecosystem, such as a forest, the pumping action of the plants ensures that a high proportion of each is retained in the system. During the plants' period of active growth, the elements are absorbed into the roots, transported to the stems and incorporated into the newly formed tissue, much of which is leaf tissue. In autumn, the leaves fall to the ground and decompose, during which process the elements are released into the soil and thus become available for absorption again in the spring. A small proportion of each element is lost with drainage water, but this is made good from the rain, by the slow weathering of soil constituents and, in the case of nitrogen, by the activities of micro-organisms in the soil which 'fix' this gas from the atmosphere and convert it into nitrate that can be absorbed by plants.

A proportion of the elements absorbed by the plants is diverted into the bodies of plant-eating animals and, through them, taken up by predators. All the material consumed by animals is returned to the soil in the course of time, either as waste products, particularly faeces and urine, or in the form of corpses when the creatures die. Faeces of large animals are a source of food for many soil-dwelling creatures, such as dung-beetles, various other insects, and slugs, and these animals redistribute the material throughout the soil. Carrion-feeders similarly redistribute the constituents of corpses.

In forests, a large proportion of each of the nutrients absorbed by the plants is returned to the soil at leaf fall; the proportion following the diversion through animals' bodies is relatively small. In grazed grasslands, however, the reverse is the case, many of the nutrients being returned rapidly in the dung of the grazing mammals. Only about ten per cent of the food consumed by sheep and cattle is retained in their bodies and most of this is removed from the grassland system when they are taken away. In hay and arable systems, most of the nutrients absorbed by the plants are removed with the crop and, if these are not replaced, a steady depletion occurs.

Circulation of water

Measurements have shown that, of the rain falling on beech forest in summer, more than 20 per cent is intercepted by the leaf canopy and evaporates again without reaching the soil; the rest of the water is discharged on to the soil but, compared with unprotected ground, the force of the impact is reduced. Partly as a result of this and partly because of the absence of trampling, woodland soil is less compact than pasture soil; indeed, the former can absorb water 50 times faster.

In summer, all the water that reaches the forest floor is returned to the atmosphere by transpiration, and the vegetation even draws on reserves stored from the winter. In winter, however, some of the water escapes from

Floor of a beech wood. A moss-covered beech-stump undergoes decomposition by a fungus.

the forest system through drainage and carries in solution minerals dissolved while passing through the soil. The evidence indicates that, when certain areas were deforested by Neolithic and Bronze Age peoples, the upper layers of the soil deteriorated badly. These layers became bleached, an indication that material had been leached from the upper layers. Some of it was deposited again lower down, in the form of hard concretions, but this material was not available to plants.

Soil types

Soils in which the upper layers have been bleached through leaching of minerals are known as 'podsols'; they often occur beneath heath vegetation and also coniferous forest. Under broadleaved forest, however, there is usually no bleached layer and the soil is uniformly brown: these soils are known as 'brown earths'.

Brown earths differ from podsols in a number of other ways. They lack the thick layer of organic matter which lies on the surface of podsols and also, because the upper layers have not been severely leached, there are no concretions in the lower. Probably because podsols are deficient in calcium, they usually hold no earthworms and are more acid than brown earths.

Agricultural land

The soils of the Northern Lowlands are derived largely from water-borne deposits laid down during the retreat of the great ice sheet which extended from Scandinavia to the northern slopes of the Central Uplands. Much of this ground is sandy, and partially podsolized, so that the nutrient-supplying capacity is low. Better quality soils occur where the silt and clay content of the parent material is higher. Because the land is so flat, water tended to build up over large areas after the retreat of the ice and, during the subsequent period of climatic amelioration, peat began to accumulate there. In the natural state, this ground is unsuitable for agriculture, but after the area has been drained, and clay and sand mixed with the surface layers, crops can be grown. The boulder clays of the moraine country south of the Baltic give rise to relatively good soils.

The Central Uplands were not glaciated and so the soils are derived directly from the underlying bedrock: the quality of the soil depends upon the type of rock. This is also true of the whole of northern France, except for those parts, particularly in Normandy, which are covered with wind-blown material known as loess, or in France as *limon*. A broad band of loess also occurs along the south side of the Northern Lowlands, in Germany. This deposit of fine, yellow material gives rise to good agricultural soils.

Climate affects agricultural activity, both directly and through the soil. Thus, where the climate is moist, clay soils tend to become waterlogged and unsuitable for arable crops, whereas sandy soils drain freely. In areas exposed to rain-bearing westerly winds, as on the Cotentin peninsula in Normandy and in the Netherlands, much of the land is under permanent pasture; the western parts of Britain and Ireland are so damp that many sections have never been cultivated and are used as rough grazing.

It is not difficult to maintain the fertility of a free-draining soil with artificial fertilizers, but these have become available in large quantities only during the last few decades. Before that, grazing of the aftermath of crops helped to maintain the nutrient status of the soils, but animals cannot put back more than they consume. Maintaining the nitrogen status was particularly difficult because, apart from what was removed by the crop, nitrate is readily leached from the soil. To recharge the soil with nitrogen, it was necessary at intervals to leave it fallow: the nitrogen status was then raised by the activities of nitrogen-fixing bacteria.

Stability of vegetation

The vegetation of arable land is unstable. Crops have to be sown every year if the land is to remain productive; they must also be tended, or they will be smothered by the wild herbaceous plants which rapidly establish themselves. The wild plants constitute the pioneer stage in a sequence of vegetation types which would culminate in closed forests; intermediate stages would be represented by scrub. Grassland, too, is unstable because, if grazing or haymaking ceases, shrubs establish themselves and the succession towards forest gets under way. Gradually, as any succession proceeds, the rate of change is reduced. The ultimate stage of forest is stable because it is self-perpetuating. Only felling, heavy grazing (which results in the destruction of young trees and therefore prevents the forest from regenerating) or significant changes of climate can then bring about a major change in the vegetation.

Structure of broadleaved forest

The plants of broadleaved forest can be separated into several different height classes, and the vegetation is then said to be 'layered'. Trees form the uppermost layer and, because they have such a marked effect on the environment, particularly in the amount of light they allow and on the humidity of the atmosphere, they may be described as the dominant species. On the European mainland, especially at damp streamsides where the trees may reach a height of 30 metres, as many as six layers can be distinguished – two each of trees and shrubs and one each of herbs and mosses. In less favourable sites, however, the number of layers is reduced. For example, some woods of sessile oaks *Quercus petraea* on acid soils in the west of Britain have only three: beneath the tree layer is one of herbs and dwarf shrubs, such as heather *Calluna vulgaris* and bilberry *Vaccinium myrtillus*, and, under that, a well-developed moss layer.

Variations in the amounts of light penetrating the upper layers of the forest considerably influence the distributions of the components of the lower; the moisture regime and nutrient status of the soil also exert marked effects. Thus, such plants as dog's mercury *Mercurialis perennis*, sanicle *Sanicula europaea*, enchanter's-nightshade *Circaea lutetiana* and wood melick *Melica uniflora* are characteristic of heavier, nutrient-rich loams or even of calcareous soils, while wood anemone *Anemone nemorosa*, bluebell *Endymion non-scriptus* and wood millet *Milium effusum* occur mostly on light and medium loams. Wood-sorrel *Oxalis acetosella*, foxglove *Digitalis purpurea* and common cow-wheat *Melampyrum pratense* can be found on the most acid woodland soils. This last species is one of a group of partial parasites belonging to the Scrophulariaceae: the rather short roots of the cow-wheat become attached to those of neighbouring grasses and withdraw water and, possibly, mineral nutrients from them.

Thus, the distribution of woodland herbaceous plants is patchy and dependent upon light, water supply and soil nutrients. Certain species sometimes form locally pure communities, but more commonly several grow together. In some of these mixed communities, there is evidence that the roots of the different species exploit different layers of the soil. This, of course, reduces competition and may be the reason why they are able to grow together.

Ferns and horsetails

The majority of woodland ferns have a short, thick, slow-growing stem, from the top of which a whorl of fronds is formed each spring. Roots grow out from the stem at intervals and supply the plant with water. Minute spores are produced in fruiting bodies, known as sporangia, on the fronds and are shed in autumn. From a spore there develops a thin, flat, kidney-shaped plant, about one centimetre across, known as a prothallus. It is on the prothallus that sexual recombination takes place, the sperm swimming across it to fertilize the ovum, from which the familiar fern grows. The prothallus has no true roots and, because it is so thin, is susceptible to desiccation: it is likely to survive only in damp conditions and, therefore, ferns are found usually in the wetter parts of the forest.

Instead of a short, upright stem which branches only occasionally, bracken *Pteridium aquilinum* has long, underground rhizomes which branch frequently. Thus, once established, a bracken plant can go on spreading indefinitely, not only throughout woodland, where it is seldom controlled, but into surrounding grassland, in time dominating large areas.

The horsetails also spread by underground rhizomes, each of which produces one or more aerial branches annually, but most of them, including *Equisetum sylvaticum* and *E. telmateia*, grow in damp conditions.

Bryophytes

Mosses and liverworts do not have true roots. Root-like filaments grow out from the stem, but these serve merely to anchor the plant. Water and minerals are absorbed from water films directly into the leaves. Rain passing through the tree canopy leaches mineral elements from the tree leaves and these augment the supply to the bryophytes beneath.

Many bryophytes are very shade-tolerant and can be found growing in dense parts of woodlands which are free from leaf litter. Where a herbaceous layer is well developed, however, the cover of bryophytes is usually sparse. They grow best in acid woodlands where herbs are scanty.

Fungi

Fungi have an important part to play in forests because they and bacteria are the main agents of decomposition. The normal state of a fungus is a mass of filaments, the mycelium, which ramifies extensively through the litter layer and the soil. At certain times of the year – especially for many species in the autumn – part of the mycelium grows to form a toadstool, which is the fruiting body. Most toadstools, such as the death cap *Amanita phalloides*, which grows on soil in forests of beech and oaks, consist of a stalk and a cap. Spores are produced on gills on the underside of the cap and, when ripe, they fall and are distributed by the wind. The spores of the stinkhorn *Phallus impudicus*, however, are immersed in a sticky substance attractive to flies, which act as the agents of dispersal.

While many fungi live on dead organic material such as wood, which they help to decompose, others such as the honey fungus *Armillaria mellea* and the beefsteak fungus *Fistulina hepatica* parasitize living plants. The latter

grows largely on trunks of oaks, but the former parasitizes many different woody species, usually with fatal results. The mycelium of many species of fungus forms an intimate association with the roots of trees, the combination of root and fungus being known as a mycorrhiza. The association is of benefit to the tree for the fungus absorbs certain minerals from the soil and passes them on to the tree.

Lichens

Lichens are a special group of fungi which have a population of completely different plants, minute single-celled algae, enclosed by the filaments of their mycelium. The algae contain the green pigment chlorophyll, so they can carry out photosynthesis to produce nutrients. The nutrients are supplied to the fungal part of the lichen by the algae, and lichens do not need to parasitize trees, or draw nutrients from dead organic matter, as other fungi do. Some grow as epiphytes on tree bark, while others are found on bare stone, such as the bright orange *Xanthoria aureola* often seen on roofs and gravestones.

Climatic range of broadleaved forest

The whole of Europe south of the Baltic, together with Britain, Ireland and the southern tip of Sweden, falls within the broadleaved, or nemoral, forest zone. The nemoral zone can be regarded as that area over which broadleaved trees are dominant, but the boundaries of this zone are not necessarily the distribution limits of the broadleaved trees. The oaks which are a major feature of so much broadleaved woodland in the Northern Lowlands extend north into southern Sweden, although there the Norway spruce *Picea abies* is the dominant species. Birches *Betula*, aspen *Populus tremula* and rowan *Sorbus aucuparia* extend into or throughout the northern coniferous, or boreal, forest without becoming dominants. North of the boreal forest and in peripheral areas of Iceland, birches form extensive low scrub. On the other hand, substantial areas of coniferous forest occur within the nemoral zone, but on the poorest soils.

The production of new leaves in the spring occurs at the expense of foods stored during the latter part of the previous season and, unless there is adequate warmth and moisture during the summer, there will not be enough of these materials to support development in the following spring. Thus, the limits of the nemoral zone are dictated by summer temperature and rainfall.

Some authorities consider that the northern boundary of the nemoral forest occurs where there is a minimum of 120 days at a mean daily temperature of 10°C or more, while others equate its minimum requirements with a

Mixed beech and spruce forest, southern Germany. The dampness in such a dense forest encourages the growth of epiphytes on the surface of a beech. Lichens are in the drier atmosphere of the upper branches, mosses on the moist trunk and ferns in the crotches where water collects.

mean July temperature of 13°C. The trees of the nemoral forest also need adequate rainfall during the growing period. Between April and November, this exceeds 25 cm over the whole of western, central and eastern Europe, and is greater than or equal to the winter rainfall. As the summer rainfall declines, the nemoral zone gives way to Mediterranean scrub vegetation in the south, and to steppe in the southeast in Rumania and the Ukraine.

At all times of the year the mean temperature declines with increasing height above sea level. In central and southern Germany, the dominant species of the broadleaved forest is the beech, but it is unable to survive at altitudes over about 1500 m. In Britain, the beech is thought not to be a natural component of high forest for more than a few miles north of the Thames valley. The altitude limit of broadleaved woodland is marked by the distribution of oaks, which are rarely common above 300 m. In the Scottish glens, however, it seems that the limit is imposed, not by climatic factors, but by the altitudes attained by suitable soils. The ranges of oaks and Scots pine *Pinus sylvestris* overlap considerably, the oaks occupying the richer soils with relatively high clay content and the pines the acid soils derived from sand, gravel or drained peat.

The shedding of leaves in autumn, induced by decreasing day length, is an adaptation that enables broadleaved trees to survive the cold season. The thin leaves are not resistant to prolonged low temperatures or the desiccating effect of biting winds. Even twigs and buds can be damaged by a spell of cold weather in early autumn, but, as winter approaches, they become dormant and by January those of oak will survive temperatures as low as −30°C. Once dormant, the buds of woody species cannot be induced to grow until they have been exposed to prolonged low temperatures; the onset of growth is then triggered by a period of suitably warm weather.

Management of broadleaved forest

Although many fragments of broadleaved forest still exist in Europe, it would be wrong to assume that any of them bears a close resemblance to the natural state. Ever since man began herding cattle or sheep and cultivating crops, he has been using the forest as a source of timber and of food for his animals. A great many of the areas where broadleaved trees once grew, and perhaps most of those in Britain, have been and are still being planted with conifers. The richness of the flora of a piece of planted woodland depends on a number of factors, especially the time that has elapsed between felling and replanting, the use to which the land was put in the interval after felling, and the status of the pre-existing woodland. When new woodland is planted soon after clear felling and the land has not been used in the meantime, most of the characteristic herbs are likely to survive. When trees are planted on arable, however, a long time elapses before the full complement of herbs is able to re-establish itself – if it ever does. Since only the trees which are useful to man are planted, it is even longer before the typical tree species return.

For many centuries, as we have already seen, the occupants of a village had certain rights in relation to any adjacent scrub or woodland, including the use of the land as a source of food for domestic animals and the taking of wood. Cattle were driven into the woodland to graze the herbs and browse the shrubs, and pigs were allowed to feed on the harvest of acorns and beech-mast. Such uses seriously affected the structure of the forest, for they tended to make it more open and also to reduce regeneration. Young trees and shrubs are destroyed by grazing and so, when the old ones die, there are none to replace them.

Dead wood, the habitat of many different plants and animals in natural forest, was taken for fuel, while certain shrubs and trees, especially hazel *Corylus avellana* and hornbeam *Carpinus betulus*, were used to supply stakes for hurdles, hop poles and the repairing of tools and agricultural equipment. When shrubs such as hazel and hornbeam are cut just above ground level, they respond by producing new shoots from the stump; this treatment, known as 'coppicing', can be repeated at intervals (the frequency depending on the rate of growth of the new shoots). Coppicing was usually carried out on a rotational basis, and animals were excluded from a recently coppiced section until the new shoots had developed sufficiently to be resistant to browsing. To avoid the necessity of excluding animals, trees were sometimes cut at a height of about two metres above the ground, a treatment known as 'pollarding'.

A form of woodland management which appears to have originated in France, and which became popular in Britain and Germany in the 15th century, was that of producing coppice-with-standards. Certain selected trees, particularly oaks, were isolated and allowed to grow naturally as the 'standards'. Because they were freed of competition from other trees, the standards produced broad crowns with thick lower branches. From the junctions of these branches with the trunk, 'knees' could be cut which were very valuable in ship-building and roof-construction.

Dominants of broadleaved forest

Estimates since the late 1950s indicate that, of the woodlands of France and England, 68 per cent and 58 per cent are dominated by broadleaved trees, while the corresponding figures for Germany and Poland are 30 per cent and a mere 12 per cent. In all countries, however, the balance is changing steadily in favour of conifers, because they grow faster and the timber they produce is more suited to modern needs. The finest remaining stand of broadleaved trees is the 5000 hectáres of the Białowieza Forest on the Russo–Polish border. Once a royal hunting ground, this forest is now a national park and the home of the only herds of European bison *Bison bonasus* living wild. The vegetation varies from alder swamp through a range of broadleaved and mixed forest types to almost pure coniferous forest. Some of the trees are several hundred years old and more than 50 m high.

European bison in the Bialowieza Forest, Poland.

The main dominants of the broadleaved forest are pedunculate oak *Quercus robur*, sessile oak and beech. In general, beeches require a moist atmosphere, annual rainfall in excess of 60 cm and, like sessile oaks, well-drained soils, but pedunculate oaks grow on heavy soils, even those which are waterlogged for short periods, and in relatively dry climates. On sandy soils in the eastern part of the Northern Lowlands, oaks are often co-dominant with Scots pines, forming the northeast mixed forest. On very wet, rich soils, alder *Alnus glutinosa* is often dominant, but the extent of this type of forest is now very restricted.

Beech forests

Provided that the climate is sufficiently moist and warm, beeches are able to grow on a wide range of sites, from the very shallow soils on the steep slopes of limestone hills to the deep, acid soils that develop from sandy ground. Beech woods still occur on and south of the Chilterns in England, on the undulating country of the ground moraine (material deposited during a temporary halt in the retreat of the ice sheet) in northern Germany and Poland, and in the moister areas of northern and northeastern France. They also occur on the hills of central Germany, at the western and eastern ends of Czechoslovakia and in the foothills of the Carpathians in southern Poland. Usually, beeches are mixed with other trees, but on the steep slopes of limestone hills they sometimes form woods on their own; in southern England, such woods are sometimes known as beech 'hangers', an example being the famous wood above Selborne in Hampshire. In extreme cases, on the shallowest soils, shrubs are absent and the herb layer is represented only by

Spring and autumn in Epping Forest. LEFT *The floor around the pollarded beeches is still thick with last autumn's leaves.* RIGHT *Mosses tend to grow where leaf litter does not accumulate.*

a few saprophytes, such as the bird's-nest orchid *Neottia nidus-avis*. Saprophytes obtain their nourishment from dead organic matter, in this case decomposing beech leaves. The vegetative parts of these plants are almost colourless, since they lack the green pigment which, in other plants, is needed for photosynthesis. Green herbaceous plants may be absent even from the lighter parts of these woods and it appears to be the dryness of the soil, rather than the low illumination, that is responsible for this.

With increasing soil depths, other herbaceous plants appear: first wood melick and broad-leaved helleborine *Epipactis helleborine*, then shade-tolerant species which require moister conditions, such as dog's mercury, sanicle, woodruff *Galium odoratum*, and common Solomon's-seal *Polygonatum multiflorum*. The climbing ivy *Hedera helix* tolerates shade, but other woody plants, such as elder *Sambucus nigra*, spindle *Euonymus europaeus*, hazel and dogwood *Cornus sanguinea*, are usually found only in the better-lit parts. In general, mosses and liverworts are scarce, mainly because the soil is too dry; they are usually most frequent near trees, where the soil receives extra water that has flowed down the trunks.

The beech woods of sandy soils, which are distinctly acid, contain pedunculate oak and also downy birch *Betula pubescens*. Herbaceous plants characteristic of the woods on neutral or calcareous soils are replaced by those of acid soils, including bilberry, heather, common cow-wheat, wood-sorrel and May lily *Maianthemum bifolium*. The moss flora is also relatively rich. A feature of such woods in Britain is that at one time the trees were pollarded: good examples are Burnham Beeches in Buckinghamshire and Epping Forest in Essex. This form of management has now been abandoned, however, and the trees have stout trunks with a characteristic crown of tall, thick branches providing, at their bases, an important microhabitat for insects.

Oak–hornbeam forests

If the rainfall is under 60 cm, or the water-table is high, the competitive ability of beech is reduced and oaks become dominant. Hornbeam, small-leaved lime *Tilia cordata*, Norway maple *Acer platanoides*, sycamore *A. pseudoplatanus*, wych elm *Ulmus glabra*, fluttering elm *U. laevis*, wild cherry *Prunus avium* and rowan are common associates. The shrubs characteristic of beech forest on deep soils, often form a well-developed layer in this type of forest, but the nature of the herb layer varies with the quality of the soil. On the better soils, tuberous corydalis *Corydalis solida*, woodruff, hedge woundwort *Stachys sylvatica*, false brome *Brachypodium sylvaticum* and a cock's-foot *Dactylis polygama* (which is softer than the commoner *D. glomerata* and has a looser inflorescence) are characteristic; on poorer, damp sites, lesser celandine *Ranunculus ficaria*, giant fescue *Festuca gigantea*, enchanter's-nightshade and greater stitchwort *Stellaria holostea* occur.

Woodlands of this type are also found in northeastern France and southeastern England. In the Netherlands, certain examples are considered to belong to this group, although they do not contain hornbeam, and it is possible that this applies also to some of the English oak woods which lie outside the range of this tree.

Carr woodlands

In low-lying areas where, sometimes, the water-table rises until it is above the surface of the soil, a luxuriant type of forest occurs in which alder, ash, Norway maple, wych elm, wild cherry and occasional hornbeams form the tree layer. The shrub layer consists largely of young trees, and the herb layer is rich in tall-growing plants, including common (stinging) nettle *Urtica dioica*, thistles *Carduus*/*Cirsium* and such ferns as *Athyrium filix-femina*. The hop *Humulus lupulus* is also characteristic. Where the land is occasionally flooded with silt-bearing water from a nearby river or stream, however, willows and poplars are the dominant trees. These types of woodland are now restricted in area because the land has been reclaimed for agriculture, but they contain many species less commonly seen and for this reason are often attractive to the botanist. In alder–ash woodlands of western Europe, for instance, yellow archangel *Lamiastrum galeobdolon*, goldilocks buttercup *Ranunculus auricomus*, star-of-Bethlehem *Ornithogalum umbellatum*, herb-Paris *Paris quadrifolia*, wood-sedge *Carex sylvatica*, black currant *Ribes nigrum* and many others can be found.

Dry oak woods

Central parts of Poland are low-lying and, because they are remote from the coast, dry. Here, fragments of a type of forest are found, in which sessile and pedunculate oaks are the dominants and the herb layer is characterized by many species which normally grow in dry habitats. The shrub layer is well-developed and includes rowan, hawthorn *Crataegus monogyna*, blackthorn *Prunus spinosa*, buckthorn *Rhamnus catharticus* and hazel. This type of vegetation occurs on south-facing hillsides.

Oak–pine and oak–birch forests

In the eastern part of the Northern Lowlands, oaks (particularly sessile oak) and Scots pine, together with silver birch *Betula pendula*, aspen and, sometimes, small-leaved lime, form a type of forest that grows on deep, sandy, podsolized soils. The shrub layer includes guelder-rose *Viburnum opulus* and alder buckthorn *Frangula alnus*, as well as rowan, hawthorn and hazel, and the ground flora consists of species characteristic of acid soils, including bracken, bilberry, May lily and also lily-of-the-valley *Convallaria majalis*.

Similar types of woodland are found in northern Germany and the Netherlands, but the amount of pine decreases with the more oceanic climate. There, too, the water-table tends to be much higher and silver birch is replaced by downy birch. On sandy, non-podsolized soils of northwestern Germany and the Low Countries, holly *Ilex aquifolium* is common. It is intolerant of winter cold and does not occur in forest east of a line from Hamburg southwards.

Woodlands of Britain and Ireland

The great majority of British and Irish woodlands have been planted. The most common dominants are the oaks, though in parts of southeastern England the hornbeam achieves equality with them and in the English Midlands the small-leaved lime is sometimes even the main tree. In general, the woods on the heavy lowland soils are dominated by the pedunculate oak, and those on the lighter upland soils by the sessile oak. Varying amounts of ash, wych elm, birch and field maple *Acer campestre* are usually present, and hazel is probably the most common shrub, but there are many woods where one of the oaks is almost the only tree species, particularly in western Britain.

In the oceanic climate of western Britain and Ireland, a rich and well-developed bryophyte layer is usually present. Extreme examples are found in southwestern Ireland: here mosses and liverworts grow as epiphytes high up in the tree canopy and form a thick mat, in which other plants, even shrub seedlings, can become established. Holly is abundant in these woods, and the strawberry tree *Arbutus unedo*, a native of Mediterranean countries, commonly occurs around the fringes. Downy birch and rowan are the main tree associates, but alder, common in woods elsewhere in western districts, is not recorded from these Irish woods. In the Scottish Highlands, birch is often the commonest tree, most of the birch woods having sprung up after the felling of former oak woods.

On the limestones of northern England, particularly on the sides of steep valleys, woods dominated by ash are frequent. Many of these woods are less than 200 years old and probably began to develop when the land was abandoned as a result of the agricultural slump following the Napoleonic wars. At about the same time the demand for timber from the mining industry was also declining. Because ash is an oceanic species, woodlands of this type are not found in central and eastern Europe.

Decomposition of litter

Every year, more than 60 per cent of the mineral nutrients absorbed from the soil by tree roots are returned in the form of litter. Nearly two-thirds of this litter is deposited in the autumn as dead leaves, and the rest throughout the year as twigs, branches, bud scales, flowers and the faeces of animals that feed upon the trees. In turn, the litter serves as food for a variety of animals that live on the forest floor and they, together with the microflora, ensure that the material is decomposed: the nutrients are then ready for absorption again the following spring.

A large proportion of each mineral nutrient is released as a result of the activities of bacteria and fungi, but, by breaking up the material into small fragments, the animals help penetration of the litter by micro-organisms. The population of earthworms (Lumbricidae) in some forest soils is capable of consuming more than 90 per cent of the leaf litter, but only a small proportion is assimilated into the body tissues. Some species, such as the garden worm *Lumbricus terrestris*, are capable of pulling leaves into their

Carr woodland in Breckland, Norfolk. Alder and wild cherry (in flower) are surrounded by a luxuriant herb layer rich in species.

burrows, but most feed by tunnelling through the litter or soil: for example, the chestnut worm *L. castaneus* lives in the litter layer, while the long worm *Allolobophora longa* and nocturnal worm *A. nocturna* occupy the mineral soil. Soil bacteria proliferate freely in the gut of a worm and it has been shown that the numbers in worm casts are far greater than in the surrounding soil. Some species, such as the long worm and the nocturnal worm, make their casts on the soil surface, but most leave them in the soil.

Millipedes (Diplopoda) and woodlice (Isopoda) come next in importance to earthworms in fragmenting litter. Both are intolerant of dry atmospheres and spend the day beneath logs, within the litter layer or wherever the air remains moist.

The animals which contribute most to the fragmentation of wood are the beetles (Coleoptera). The larvae of a click-beetle *Ctenicera pectinicornis*, a wood-boring beetle *Dicerca berolinensis*, a cardinal-beetle *Pyrochroa coccinea*, the stag beetle *Lucanus cervus*, the lesser stag beetle *Dorcus parallelipipedus* and the musk beetle *Aromia moschata* all live in rotting logs, but it is not always clear to what extent they derive nourishment from the woody material itself. Relatively little is known about the feeding habits of such beetles, but they probably rely to a significant extent on other animals and fungi inhabiting the wood.

Plant material consists largely of cellulose, the basic constituent of the cell walls, and lignin, which thickens the cell walls to form wood. Most of the

Dry oak–birch forest, Yorkshire. The low shrub layer of bilberry and absence of young trees are the results of grazing by sheep.

animals that feed on plant litter do not themselves produce the enzymes necessary to digest these materials and have to rely on the bacteria within their guts to do it for them. One group of animals that does have the necessary enzymes is the molluscs. Slugs and snails are very common in many types of woodland and most of the species feed on decaying leaves. Rounded snails *Discus rotundatus* and plaited door snails *Cochlodina laminata* both feed on wood, however, and the former also frequently eat the fruiting bodies of fungi. The netted slug *Agriolimax reticulatus* consumes large quantities of living herbaceous plant material.

Molluscs, particularly snails, require considerable quantities of calcium in their diet for the construction of their shells. Indeed, the two most important environmental factors affecting their distribution are the nature of the ground and the moisture regime. In woods on limestone, the large black or red slug *Arion ater* and many snails, including the rounded, plaited door, grove *Cepaea nemoralis*, smooth glass *Aegopinella nitidula*, clear glass *A. pura*, edible *Helix pomatia*, slippery *Cochlicopa lubrica*, crystal *Vitrea crystallina* and hairy *Trichia hispida*, may all be common. As the ground becomes more acid, however, the number of species declines until only a very few species, such as the hollowed glass snail *Zonitoides excavatus*, may be present on the most acid soils. It appears that the distribution of plant species has little effect on the distribution of snails and slugs, for the majority will eat almost any decomposing vegetation.

Despite their size, molluscs as a whole consume only relatively small amounts of leaf litter. A more significant part is played by the larvae of flies (Diptera), particularly of crane-flies and St Mark's fly *Bibio marci*, both of which feed directly on leaf mould. Many other species found in the litter probably feed on other animals or fungi.

Forest-floor predators

To the extent that they prey on animals that break up plant material, predators can be said to reduce the rate at which nutrients are made available to plants. Common arthropod predators of the forest floor are ground-beetles, such as *Harpalus rufipes*, and the centipedes (Chilopoda). Centipedes have one pair of legs on each body segment (whereas millipedes have two) and the appendages of the first segment are modified into poison claws, with which the animal catches and paralyses or kills its prey. Unlike insects, which have a waxy outer covering or cuticle, centipedes, millipedes and woodlice are not protected against water loss. They are obliged to live in moist places, spending the day beneath logs, within the litter layer, or wherever the air remains humid.

Plant-eating insects

Although many different insects feed on living plant material, the amount consumed is usually small in comparison with a season's total production.

Exceptionally, though apparently more often in the past, swarms of a cockchafer beetle, or maybug *Melolontha melolontha* attack the canopy of oaks, stripping the foliage in a single day. More regular defoliators are the larvae of several small moths such as the winter moth *Operophtera brumata*, whose females are wingless, and the green tortrix *Tortrix viridana*. Heavy infestations of larvae provide a bonanza for small, insectivorous, woodland birds, some of which nest in holes. They are encouraged by foresters, especially in Germany, by the provision of nestboxes.

As a group, the moths and butterflies (Lepidoptera) are plant-eaters. Many of their larvae that feed on leaves consume all the tissues, and the material that they eject as faeces drops down and soon becomes part of the soil again. Others, such as those of the leaf-miner moth *Lyonetia clerkella*, feed inside the leaf, consuming only the soft inner tissues, so their faeces remain within the leaf and are not returned to the soil until autumn. Many of the bugs (Hemiptera) also feed directly on plants. These insects have piercing mouthparts, which can penetrate to the conducting tissues of the leaf or twig. They feed on sap, and their waste products are rapidly returned to the soil.

In Britain, more than 180 species of butterflies and moths feed on oaks and over 170 on birches, though there are not more than 40 on ash, beech, field maple and hornbeam. More than 50 species of bugs feed upon oaks, but less than ten on beech, field maple and hornbeam. Oaks and birches are believed to have been part of the British flora for much longer than these other trees, and it is generally the case that the more recently introduced a plant species is, the smaller the number of insects which feed upon it.

Woodland is the true home of many brightly coloured butterflies and the majority of the fritillaries are to be found there. The caterpillars of some species, such as the purple emperor *Apatura iris*, the poplar admiral *Ladoga populi* and the hairstreaks *Quercusia quercus* and *Nordmannia ilicis*, feed on trees, but those of many others, including the fritillaries, are found on herbaceous plants. Although butterflies are common in the woodlands of the European mainland, they are relatively scarce in Britain and mostly limited to the south of England, probably because of the climate. Many woodland species occur in glades and along rides where they find flowers on which to feed, but the wood white *Leptidea sinapis*, speckled wood *Pararge aegeria* and woodland brown *Lopinga achine* are often in deep shade. Recently, however, it has been shown that the speckled wood occupies a 'travelling territory' as a patch of sunlight moves through the wood during the day. Most butterflies take nectar, but several species consume the honeydew from aphids, while the purple emperor will also feed on carrion and the sap oozing from tree wounds. In fact, a carrion bait is used by collectors to attract the normally high-flying purple emperor; it will also come down to land on shiny surfaces such as the body of a motor-car.

Among the most interesting insects that feed on plants are those that stimulate the abnormal growths known as galls; these include the gall-wasps

Elm trees in a hedgerow showing the effect of Dutch elm disease.

of the family Cynipidae (Hymenoptera). When a gall-wasp lays an egg, it injects into the plant a substance similar to the plant's own growth-hormone and this is responsible for the development of the gall. Some galls, such as the oak-apple stimulated by the gall-wasp *Biorrhiza pallida*, hold many other species of larvae as well: these include predators, parasites and larvae that feed upon the gall tissue. More galls are formed on oaks than any other plants.

Even the tough, woody tissue of trees is not immune to attack by insects. Wood is not a very nutritious material and this may be why some larvae, such as those of the goat moth *Cossus cossus*, take a long time to develop. This insect normally spends three or more years as a larva, but, when fed artificially on other plant materials, it can complete its development in one year. The tongue of the adult goat moth is only rudimentary, and it does not feed at all.

Particularly important as wood-borers are the bark-beetles. Each female bores through the protective bark and then excavates a tunnel in the sap wood. She lays several eggs in this main gallery and, on hatching, the larvae make their own tunnels, usually at right angles to the main gallery. The adult beetle eventually escapes by boring its way out through the bark. The elm-bark beetles *Scolytus scolytus* and *S. multistriatus* feed on elm wood and, if the tree happens to be infected with the fungus *Ceratostomella ulmi*, which is responsible for Dutch elm disease, the escaping adults are likely to carry some of the spores with them. The beetles fly to the young shoots of elms to feed and so may deposit spores on the damaged tissues of uninfected trees. The fungus, now rampant in Europe, grows rapidly and kills the trees by

LEFT *Large elm-bark beetles on the bark of a dying elm. The sawdust is from their tunnels excavated under the bark.* RIGHT *Speckled wood butterfly in a sun-fleck resting on a beech seedling; New Forest.*

causing the blockage of the passages through which water flows to the branches.

Insects and spiders as predators

The rapid cycling of nutrients by plant-eating insects is to some extent held in check by predators and parasites. Aphids are eaten by a variety of insects, including such predatory bugs as the thread-legged bug *Empicoris vagabundus*, elm gall-bug *Anthocoris gallarum-ulmi* and oak-bug *Cyllecoris histrionicus*, as well as by the adults and larvae of ladybirds (Coccinellidae) and the larvae of many hover-flies (Syrphidae). When leaves of wych elm are attacked by the aphid *Eriosoma ulmi*, they tend to curl and become somewhat deformed. This aphid lives within the rolled leaf, as does its predator, the elm gall-bug. The oak-bug frequently feeds on the larvae of the gall-wasp *Neuroterus quercus-baccarum*, which form spangle galls on the leaves of oak.

Insects on the wing are prone to attack by robber-flies, such as *Leptogaster cylindrica*, and by wasps which take them as food for their grubs. Flying insects are also liable to be ensnared by web-building spiders: two common types of web found in woodlands are the vertically orientated orb-webs built by such species as *Tetragnatha montana*, *Meta segmentata* and *Araneus umbraticus*, and the horizontal hammock-webs constructed by *Linyphia triangularis*. At least some of the strands of orb-webs are sticky and hold the insect till the spider reaches it. The hammock-web of *L. triangularis* is, however, not sticky; it is supported above and below by many oblique strands and, if an insect collides with one of these it may fall on to the hammock. This spider bites and paralyses its prey from below, then cuts a hole in the hammock, drags the insect through and wraps it with silk threads.

Herbivorous mammals

Although several mammals contribute to the rapid cycling of nutrients by consuming vegetation, not all of them feed entirely on vegetable matter. The wild boar *Sus scrofa* and the badger *Meles meles* eat a variety of animal foods including earthworms, molluscs, rodents and young rabbits, while the hazel dormouse *Muscardinus avellanarius*, garden dormouse *Eliomys quercinus* and fat dormouse *Glis glis* sometimes eat insects or small birds. The voles amongst the small mammals and the deer amongst the large are almost entirely herbivorous.

The wood mouse *Apodemus sylvaticus*, bank vole *Clethrionomys glareolus* and dormice move freely along the branches of shrubs from which they collect berries and nuts in the autumn. They all store food for the winter but only the dormice enter a state of true hibernation with reduction of body temperature and metabolic rate. While these animals may be harmful to certain species of trees and shrubs if they consume enough of the seeds to affect their reproductive potential seriously, they may also act as dispersal agents because seeds stored but not eaten are likely to germinate and develop a long way from the parent plants.

The indigenous European red deer *Cervus elaphus*, fallow deer *C. dama*, and roe deer *Capreolus capreolus*, and the introduced sika *Cervus nippon* from Asia, are essentially animals of broadleaved forest, but in Scotland, following the 1745 rebellion, many of the deer forests were cut down by order, and the red deer established itself on moorland. During the summer it is frequently found on the summits of the highest mountains. The fallow deer is a native of southern Europe but it has been widely introduced into countries of central Europe, Britain and Ireland, southern Sweden and southern Norway. These animals are both grazers of herbs and grasses and browsers of shrubs, their diet varying according to season and opportunity, and also from one population to another. Deer can do considerable damage to young trees by using them as rubbing posts to remove the velvet from their hardened antlers and in severe winters they may strip and eat bark from young trees.

Despite the large-scale destruction of its habitat, broadleaved forests, and the fact that it is hunted, the wild boar is still widespread across the lowlands of Europe. Its fecundity must contribute considerably to its resilience, for there can be up to 12 young in a litter and, when food is plentiful, a sow may have two litters a year. Sows and young live together in small groups. They are nomadic and often move many kilometres in the course of a night, but daytime is spent under cover of woodland with well-developed undergrowth. They eat mostly those parts of plants with high food-value, such as large seeds, bulbs, tubers and rhizomes.

Predatory mammals

Of the predatory mammals that have survived persecution by man, the most sylvan species, except in Britain, is the pine marten *Martes martes*. The others range over the surrounding open country in search of food and use small woods or the edges of large woods as cover during the day. The pine marten, beech marten *Martes foina*, polecat *Mustela putorius* and red fox *Vulpes vulpes* all feed on other mammals, birds and sometimes amphibians, and the fox also consumes carrion.

All the European species of bats feed on insects taken in flight. The wing of a bat consists of a membrane of skin stretched between the exceptionally long digits of the forelimb, across to the hindlimb and thence to the tail. Moths and beetles figure prominently in the diet of most bats, probably because many of them, like the bats, are nocturnal. When a lesser horseshoe bat *Rhinolophus hipposideros* or a greater horseshoe bat *R. ferrumequinum* catches a large insect, it presses it against the wing membranes close to its body, bites off the inedible parts and eats the rest, flying all the time. These bats have short tails. Other, longer-tailed species direct the tail forwards and hold an insect against the membrane joining it to the hindlimb. Sometimes a bat will alight to tackle its prey, usually using the same perch; a litter of insect wings can then be found beneath. Bats are most frequently found along woodland fringes or in the neighbourhood of water, places where the numbers of insects are greatest. During the day they roost in hollow trees, under loose bark, in buildings, in caves or wherever they can find suitable cover.

Although its 'hills' are such a familiar sight in open grassland, the mole *Talpa europaea* is equally at home in closed woodland. The hedgehog *Erinaceus europaeus* and the shrews (Soricidae) occur on the fringes of woodland but mostly in more open country. The mole spends nearly all its life underground feeding on a variety of invertebrates, especially earthworms. It is active throughout the winter and in cold weather stores earthworms in its mounds; the anterior end of the worm is bitten so that it is no longer capable of co-ordinated movement and hence of escape. By consuming large numbers of soil-inhabiting invertebrates, it is probable that, on balance, the mole has a retarding effect on the rate of circulation of nutrients.

Birds

In winter the bird fauna of a large area of broadleaved forest may seem surprisingly sparse. This is partly because many of the summer residents are migratory and, in autumn, leave Europe for the Southern Hemisphere and partly because many of those that remain aggregate into mixed flocks. The flocking habit is adopted by tits *Parus*, long-tailed tits *Aegithalos caudatus*, nuthatches *Sitta europaea*, treecreepers *Certhia* and goldcrests *Regulus regulus*. Parties of these birds, occasionally accompanied by one or two smaller woodpeckers, follow regular routes, keeping together with high-pitched 'contact calls' as they search for wintering stages of invertebrates.

Where there are trees and shrubs bearing berries, thrushes *Turdus*, including fieldfares *T. pilaris* and redwings *T. iliacus*, may often be found. Chaffinches *Fringilla coelebs* and bramblings *F. montifringilla* congregate to feed on the seeds of beech, while hawfinches *Coccothraustes coccothraustes* take those of hornbeam. Flocks of woodpigeons *Columba palumbus* also feed on a wide variety of fruits and seeds including beechmast and acorns. Acorns are a favourite food of the jay *Garrulus glandarius* in winter, but other members of the crow family (Corvidae) are mostly scavengers or predators in broadleaved forests.

In spring the winter visitors depart, the residents disperse to their breeding territories and the summer visitors return. The latter include warblers (Sylviidae), flycatchers (Muscicapidae), redstarts *Phoenicurus phoenicurus*, golden orioles *Oriolus oriolus* and rollers *Coracias garrulus*. Resident populations of blackbirds *Turdus merula*, song thrushes *T. philomelos*, robins *Erithacus rubecula* and wrens *Troglodytes troglodytes* may be augmented by individuals which have spent the winter in open country or near human habitations and by migrants.

Not all of the species that breed in forests are strictly arboreal in their habits. Several of the warblers invariably build their nests on or close to the ground while others favour bramble thickets. Finches, jays, song thrushes and blackbirds usually build in the taller shrubs while the canopy layer provides nest-sites for comparatively few species: these are mostly large birds such as certain birds of prey, the carrion crow *Corvus corone* and the woodpigeon. Woodpeckers, tits, most flycatchers, some owls and the stock dove *Columba oenas*, redstart, starling *Sturnus vulgaris* and jackdaw *Corvus monedula* occupy holes in trees, but only the woodpeckers and two or three of the tits, notably the willow *Parus montanus*, actually excavate the holes themselves and then usually in rotting wood which is relatively soft. But the larger woodpeckers, the black *Dryocopus martius*, associated with beech woods, and the green *Picus viridis*, can chisel through hard outer wood before sinking a shaft in the rotten core. Their holes are attractive to other animals, such as bats, squirrels, starlings and, where there are several of the black woodpecker close together, small groups of stock doves.

Almost all the woodland birds eat invertebrates, at least during the breeding season. Several feed on fruit when it is available and some on seeds, but, in striking contrast to the insects and mammals, none feeds regularly on the leaves of plants. In general, birds probably exert a retarding influence on the rate of circulation of nutrients, but there is no firm evidence that they are able to control numbers of insects to any significant extent. In natural conditions, however, the densities of some insectivorous birds, such as tits, redstarts and hole-nesting flycatchers, are probably limited by the availability of nesting sites: their numbers can often be greatly increased by the provision of nestboxes.

The woodland habitat can be separated into a variety of microhabitats and each species utilizes only a limited number of these. Thus, woodpeckers and treecreepers spend much of their lives on tree trunks, while the woodcock *Scolopax rusticola* lives mostly on the ground. On the other hand, the willow warbler *Phylloscopus trochilus* and wood warbler *P. sibilatrix* occupy several microhabitats because, although they nest on the ground, they feed in the shrub and tree layers. The microhabitats of any species together constitute its ecological niche, but this term must be broadened to incorporate the animal's role in the woodland system. Thus the niche of a warbler which nests on the ground and feeds in the bushes is different from that of a blackbird which nests in a bush and feeds on the ground. Other factors such as type of food eaten and requirements for song posts must also be included, with the result that in the event no two species occupy precisely the same niche.

The dominant species of a broadleaved wood has such a marked effect on the structure of the wood that it is almost impossible to separate the effect of the dominant itself from that of the structure on the bird-life. A number of birds appear to have definite preferences with regard to broadleaved and coniferous species however, and where trees of both types occur birds associated with each are usually present. Thus the bird-life of a mixed forest is usually richer than that of a forest composed of broadleaved or coniferous species alone and, compared with broadleaved forest, is likely to be augmented by species such as the coal tit *Parus ater*, goldcrest and firecrest *Regulus ignicapillus*.

Forest raptors

Although the majority of European raptors (birds of prey) breed in trees or on wooded cliffs, only those that hunt in forest can be classed as forest birds. Individuals of the tawny owl *Strix aluco* and long-eared owl *Asio otus* may spend much of their lives in woodland and so may the goshawk *Accipiter gentilis*, sparrowhawk *A. nisus* and honey buzzard *Pernis apivorus*. Other tree-nesting species, such as the buzzard *Buteo buteo*, booted eagle *Hieraaetus pennatus*, short-toed eagle *Circaetus gallicus*, lesser spotted eagle *Aquila pomarina* and spotted eagle *A. clanga*, tend to hunt over open country.

Most raptors have regular roosting sites where they disgorge the indigestible parts of their prey in the form of large pellets. The food of the bird can be determined from the bones, especially skulls, and pieces of insect exoskeleton in the pellets and it has been shown that the proportions of different types of prey taken varies with the habitat in which the bird lives. Thus tawny owls living in urban environments eat large numbers of small birds, especially house sparrows *Passer domesticus*, whereas those living in woodland eat very few. The ability to modify its feeding habits must have been important among the factors enabling the tawny owl to exploit man-made habitats so successfully.

HEDGES

Origins of hedges

Until the middle of the 18th century, a large proportion of the agricultural land of Britain was farmed on the open-field system. Each farmer in a community had strips of land in each of the two to six large fields associated with the village. The system was inefficient and, to put new agricultural techniques into practice, pressure developed from the more wealthy landowners for holdings to be consolidated. Consolidation of land in a district was brought about by an Act of Parliament, and in the latter half of the 18th century many such Acts were passed. A farmer was obliged to protect his crops from a neighbour's animals, so there was a requirement that each parcel of land be enclosed. This was generally carried out by planting a hedge of hawthorn *Crataegus monogyna* and, to restrict the movements of stock and facilitate the rotation of crops, each parcel was usually subdivided by more hedges.

Even today, the majority of British hedges consist of hawthorn, though sometimes English elm *Ulmus procera* is the dominant. A small proportion of 'mixed' hedges contain several species of shrub, some of which are present in approximately equal amounts and together make up the greater part of the hedge. A number of these hedges were probably planted as a mixture, but it is significant that some are sited on the boundaries of ancient enclosures, while others occur on land known to have carried woodland within the last few centuries.

Many of the species present in mixed hedges, such as hazel *Corylus avellana*, field maple *Acer campestre* and holly *Ilex aquifolium*, commonly form an understorey in existing fragments of broadleaved woodland. Furthermore, the ground flora of these hedges frequently consists largely of herbs characteristic of woodlands, such as bluebell *Endymion non-scriptus*, wood anemone *Anemone nemorosa*, dog's mercury *Mercurialis perennis* and lords-and-ladies *Arum maculatum*, together with honeysuckle *Lonicera periclymenum*, ivy *Hedera helix* and other woodland climbers. In some areas, however, hedges are composed of unusual mixtures: thus, in parts of the flat, low-lying Plain of York, alder *Alnus glutinosa*, downy birch *Betula pubescens*, willows *Salix* and even rowan *Sorbus aucuparia* are common ingredients, probably remnants of the wet scrub and forest which at one time covered much of the area.

A hedge rarely consists entirely of a single species. In the majority of British hedges hawthorn is predominant, but there are usually a few, scattered individuals of other shrubs, especially elder *Sambucus nigra*, ash *Fraxinus excelsior* or blackthorn *Prunus spinosa*. Where a hedge extends away from a plantation of sycamore *Acer pseudoplatanus*, it is common to find sycamores in the adjacent part. The distribution of all these shrubs suggests that they have developed from seeds which, presumably, were carried there either on the wind or by animals. So, with time, the composition of a hedge gradually becomes more varied; its age can sometimes be estimated by the average number of species of shrubs in a given length. The formula expressing the relationship was determined by comparing the ages of hedges known from historical records with the number of species in a length of 30 yards (27.4 m). Great caution should, however, be exercised in applying this formula, for so far it has been shown to apply only in the southern half of England and, even there, it is unlikely to be meaningful in relation to hedges derived directly from woodland.

Distribution of hedges

Despite the extensive uprooting of hedges since the 1950s, they are still a common feature of lowland Britain. On the Continent, however, hedges are much less widespread. In Denmark, they are a feature particularly of the centre of Jutland, and the Netherlands have some particularly rich examples in the Drenthe district south of Gröningen. The latter often include shrubs characteristic of wet woodland, including alder, alder buckthorn *Frangula alnus*, buckthorn *Rhamnus catharticus*, several species of willow, downy birch and guelder-rose *Viburnum opulus*. In France, hedges are abundant on farmland in Brittany and adjacent areas; in Germany, apart from the northwest, they are rare. Even today, the majority of German farmers have small holdings of ten hectares or less and each consists of several separate pieces of land. This is the result of *Realteilung* whereby, on the death of a farmer, the land is divided between his heirs. The enclosure movement did not extend to Germany, so that the separate parts of a holding have not been consolidated and, in consequence, hedges have not been planted. The system is inefficient and since the 1939–45 war attempts at consolidation have been made.

In western France, as in Britain, boundaries were often marked by a bank, and an adjacent ditch resulting from the bank's construction. Hedges developed or were planted on the bank and, especially in the northwest, consist of shrubs and pollarded trees, usually oak *Quercus* and ash. As well as serving to control the movements of stock, these hedges are a source of fuel.

Management of hedges

The way in which hedges are managed varies considerably from one locality to another. Since one of their main functions is to keep domesticated animals in or out, they should be stock-proof. In southeast England, the management of hedges often resembles that of coppice woodland: the shrubs are periodically cut close to the ground and the stakes used on the farm. The shrubs

respond by producing new shoots and, when they have developed sufficiently to make a stock-proof barrier, the adjacent field can again be used for grazing. Elsewhere, selected stems are 'laid', so that the hedge remains stock-proof while the new growth is developing. Laying is practised particularly in the English Midlands and also in parts of the *bocage*, the area where land is enclosed by hedges in France.

With the spread of crop monoculture and intensive livestock-rearing, hedges have become superfluous on many farms and a large proportion have been removed. Thus the farmer is relieved of the financial burden of maintaining the hedges, additional land is made available for cultivation and, it is claimed, pests can be controlled more effectively. On some farms, the hedges that remain are still managed by traditional methods, but on many others cutting machinery attached to a tractor has been substituted for the farm labourer and hedges are then often reduced to a 'tidy' height of about one metre. Few animal species are found in such hedges, compared with the number in those allowed to grow unchecked.

The hedge as a habitat for animals

The flora of a species-rich hedge usually bears a considerable resemblance to that of broadleaved woodland, but it does not necessarily support many of the characteristic animals. The degrees of exposure, light and humidity within a narrow hedge which is cut regularly are so different from those in the interior of woodland that it is quite unsuitable for many species. With increasing breadth and height, however, more woodland-like conditions are produced and other animals will be able to live there. Indeed, in many ways, a broad, well-developed hedge resembles the fringe of a wood.

The flower buds of most European shrubs are formed in late summer and early autumn. Some species have them on the current year's growth, but others produce them on wood two or three years old. A hedge is usually cut in autumn or winter and, since the summer growth is removed, shrubs of the former type will not produce flowers the following season. Even those of the latter type usually also fail to flower because the amount of wood of suitable age is so small. A regularly trimmed hedge will therefore not attract visiting insects that feed on flowers and it will, of course, produce no food for fruit-eating animals.

For some animals, such as plant-feeding insects, those that live in leaf-litter, and certain small mammals, an untended hedge provides all the essentials of life: food, shelter and breeding sites. It also acts as a highway along which they can move at minimal risk from predators. Woodland birds able to find sufficient food in open country are more likely to occur where there

An untended hawthorn hedge provides an abundance of flowers attracting insects in spring, and berries for birds and mammals in the autumn. Plants such as male fern (above left) suggest that this hedge might be derived from woodland.

are tall hedges because these provide song-posts and nest-sites at a suitable height. Dead trees help to enrich the animal population of a hedge, especially if part of the trunk or some of the branches are hollow. Water-holding cavities serve as breeding sites for certain insects with aquatic larvae, and dry ones provide nest-sites for such birds as tits *Parus* and owls (Strigiformes). The potential of a hedge is also enhanced if a ditch runs alongside, for this attracts moisture-loving or even aquatic plants and animals.

Untended hedges are not favoured by arable farmers because they take up more land and are more likely to be occupied by rabbits *Oryctolagus cuniculus*, but they are more frequently tolerated around pastures because they provide shelter for stock in bad weather. Perhaps the best compromise is to allow a hedge to grow to a height of about two metres and trim it so that it is triangular in section. Such a hedge provides more ground cover than one which is rectangular but occupies the same amount of space.

Ground-dwelling invertebrates

In areas of intensive arable farming, hedges may be the only habitat in which leaf-litter is present in sufficient quantity for the animals feeding on it to survive. As in woodland, other environmental factors, particularly the humidity at ground level and the leaf material available, determine the species present. Among woodlice, for example, *Philoscia muscorum* and *Porcellio scaber* may be found in dry litter, but the widespread *Oniscus asellus* and the small *Trichoniscus pygmaeus* require rather moister conditions. The presence of suitable prey will bring into the hedge predatory ground-dwellers, such as centipedes (Chilopoda) and ground-beetles (Carabidae), and it is also likely that the carrion-beetle *Oiceoptoma thoracicum* and other carrion-feeders will occur. The greater the variety of herbaceous plants, the greater will be the range of microhabitats and hence the number of ground-dwelling animals.

Insects in hedges

Insects are selective as to the species of plant on which they feed, so one of the factors that affects their total numbers in a hedge is the variety of plants. Two of the commonest shrubs in British hedges, hawthorn and hazel, each support an exceptionally large variety of plant-eating insects. The total number exceeds 230 and 100 respectively, though in Britain the figures are only 150 and 70. The genera *Prunus* (blackthorn, plums and cherries) and *Malus* (apples) also each support many species of insects, but even those shrubs that support relatively few, such as buckthorn, dogwood *Cornus sanguinea* and guelder-rose, are important as components of a mixed hedge, because they allow insects with very specific food requirements to live there. Thus, the larvae of the brimstone butterfly *Gonepteryx rhamni*, which feed on buckthorn and alder buckthorn, are likely to occur in a hedge containing either of these species, and the larvae of the red underwing moth *Catocala nupta* where willows or poplars *Populus* are present.

Shrubs that are not regularly trimmed are likely to flower and these will attract butterflies (Lepidoptera), bees (Hymenoptera) and other insects that feed on nectar and pollen. Plants which flower early in the year, such as the willows and hazel, provide food for queen bumblebees *Bombus* which emerge earlier than most insects. Flowering plants are also likely to produce fruit in due course, thus giving sustenance to fruit-eating insects such as the hawthorn shield-bug *Acanthosoma haemorrhoidale* which feeds on haws. A rich ground flora also adds considerably to the variety of insect life. Common (stinging) nettles *Urtica dioica* are the main food-plants for the larvae of several of our most colourful butterflies, including the peacock *Inachis io*, red admiral *Vanessa atalanta*, painted lady *Cynthia cardui*, small tortoiseshell *Aglais urticae* and comma *Polygonia c-album*. Two of these, the red admiral and painted lady, are unable to survive the average winter north of the Alps, and every spring northern Europe, including Britain, is restocked by immigrants from the Mediterranean area; the new arrivals breed, and by late summer both species are usually quite common.

The plant-eating insects fall prey to a wide variety of predatory species. The larvae of ladybirds (Coccinellidae), lacewings (Chrysopidae) and several of the hover-flies (Syrphidae) all feed on aphids, but, while the ladybird larvae crush them with their powerful mandibles, those of the lacewings and hover-flies pierce the outer covering, or exoskeleton, of their victims and suck out the body fluids. Adult predatory bugs, including the elm gall-bug *Anthocoris gallarum-ulmi* and delicate apple capsid *Malacoris chlorizans*, feed in a similar way and so do robber-flies (Asilidae), such as *Leptogaster cylindrica*. This last species takes a variety of insects on the wing and probably paralyses them with a poison injected through its mouthparts.

The larvae of butterflies and moths are particularly prone to attack by parasitic flies and by ichneumon flies (Ichneumonidae). Some of these, such as the tachinid fly *Nemorilla floralis* and the ichneumon fly *Netelia testacea*, attack the caterpillars of several species. The eggs of the parasite are laid in the body of the young caterpillar and, as the larvae grow, they gradually consume its body substances; the host dies, usually just before pupation, and the larvae of the parasite then emerge and pupate. In some cases, the eggs are laid actually within the egg of the host, as happens when the small ermine moth *Yponomeuta padella* is parasitized by one chalcid wasp, a species of *Ageniaspis*. The eggs of the parasite develop in such a way that several embryonic larvae are produced which then begin to feed on the host's body tissue.

By no means all insects that frequent hedges necessarily feed there. Thus, the queens of several species of bumblebees and wasps build their nests in hedges, but the workers may feed a long way off. Some bumblebees, such as the large red-tailed *Bombus lapidarius*, have long tongues, while others have short ones. The long-tongued species are able to reach the nectar in certain

LEFT *Small tortoiseshell caterpillars feed gregariously on stinging nettles and leave behind them webbing, frass and eaten leaves.* RIGHT *An overwintering female brimstone, encrusted in ice, hangs leaf-like from a twig.*

fodder crops, including clover and alfalfa, without damaging the flowers and in so doing they pollinate them. Since the plants fail to produce seed unless they are cross-pollinated, these bumblebees are important to the farmer.

Thus, of the great variety of insects that live in a hedge, some, like aphids, are potentially harmful to crops, while others, such as bumblebees, are beneficial. Most of the plant-eating insects are subject to the attacks of predators and parasites, which, in turn, may themselves become victims of others. Any attempt to work out quantitatively the relationship between just two species – for example, the extent to which a predator is affecting the numbers of a particular plant-feeder in a given set of conditions – is clearly a formidable task.

Birds and hedges

So far as plant composition is concerned, hedges bear at least a slight resemblance to woodland, but in their shape they do not resemble any natural habitat. In terms of animal evolution hedges have been available for only a short period of time, so it is not to be expected that any species would occupy hedges to the exclusion of any other type of habitat.

Most of the birds that breed in hedges also breed in woodland. Since the latter is the natural habitat for these species, it is to be expected that the

Fruit debris in an old bird's nest used as a feeding platform by a small mammal, probably a wood mouse, which eats only the seeds and discards the flesh of the fruit.

number that adopt a hedge as a suitable habitat will depend upon the extent to which it is similar to woodland. There is some firm evidence that this is so. Thus, in one study, the number of species recorded as breeding in hawthorn hedges with outgrowths was nineteen whereas the average number in hedges regularly cut by machine was only seven. Elm hedges were included in the same study and it was found that the average numbers in the two types were fifteen and five respectively. Thus again, from the naturalist's viewpoint, it is fortunate that hawthorn was so widely used for hedges.

The results of various studies all indicate that the most common species breeding in British hedges are the robin *Erithacus rubecula*, song thrush *Turdus philomelos*, blackbird *Turdus merula*, chaffinch *Fringilla coelebs*, whitethroat *Sylvia communis*, dunnock *Prunella modularis* and yellowhammer *Emberiza citrinella*. In the study quoted above, the species breeding in the hedges with outgrowths, which were not found in any other type of hedge and which apparently require the hedge to resemble a woodland or woodland edge as closely as possible, were goldfinch *Carduelis carduelis*, blackcap *Sylvia atricapilla*, garden warbler *S. borin*, lesser whitethroat *S. curruca*, turtle dove *Streptopelia turtur* and wren *Troglodytes troglodytes*. Amongst the seven most common species, the first four also occur frequently in closed-canopy forest, and the whitethroat, dunnock and yellowhammer occur in scrub.

Of the species mentioned it could be argued that the yellowhammer is the bird most adapted to open country, because it frequently builds its nest at ground level in the hedge bottom or in rough grass. Furthermore, in parts of Europe, the type of habitat occupied in Britain by the yellowhammer is taken over by the cirl bunting *Emberiza cirlus* and the yellowhammer is then relegated to open montane areas. Other species which frequently nest on the ground in hedge bottoms, but also away from hedges, are the grey partridge *Perdix perdix*, corn bunting *Miliaria calandra* and reed bunting *Emberiza schoeniclus*. Thus the breeding bird fauna of hedges appears to consist of those species which find the habitat a reasonable substitute for woodland, together with those of open country which take advantage of the cover it provides. The reed bunting is of particular interest because it appears to be in the process of extending its habitat from wetlands into the drier types of vegetation occupied by the closely related yellowhammer.

All of the species mentioned so far could be found breeding in hedges consisting only of shrubs. The presence of trees adds another stratum but few birds nest in trees unless they are old and have developed holes in the trunk or branches or carry a thick growth of ivy. Such trees may induce other woodland species, such as tits or owls, to occupy the hedge.

Many of the shrubs which occur in hedges, including hawthorn, dogwood, alder buckthorn, buckthorn and guelder-rose, produce berries, and flocks of birds, particularly the thrushes, visit the hedges to feed upon them. Because they are fully exposed to the light, these shrubs, when allowed to grow naturally, produce much heavier crops of berries than they do in woodland.

Mammals and hedges

The commonest of the small, herbivorous mammals in British hedges are the bank vole *Clethrionomys glareolus* and the wood mouse *Apodemus sylvaticus*. Both are good climbers but it is the wood mouse which is largely responsible for all the fruit debris which is commonly found in old birds' nests in autumn. This mouse collects rose hips and hawthorn berries and often uses a bird's nest as a platform on which to feed. The wood mouse consumes the seeds but rejects the flesh of these fruits whereas the bank vole eats the flesh.

These animals burrow in the soil beneath the hedge and also make extensive systems of tunnels through the undergrowth. The bank vole spends most of its life in the hedge whereas the wood mouse makes frequent sorties into the neighbouring fields to feed. Indeed, wood mice seem to be able to live entirely in the fields.

Other common inhabitants of hedges are rabbits, which make extensive warrens especially when the hedge is situated on a bank, and shrews, both common *Sorex araneus* and pygmy *S. minutus*. Few of the woodland predators are found in hedges; only the stoat *Mustela erminea* and weasel *M. nivalis* are common here.

GRASSLAND AND SCRUB

Types of grassland

The cultivation of crops involves much time and effort whereas, unless maximum production is the aim, grasslands need relatively little attention. It is not surprising, therefore, that arable crops should be grown on the more fertile and accessible soils, while inaccessible sites with poor soils are used for grass. In times of agricultural depression poorer arable land is often allowed to 'tumble down' to grass, whereas during boom periods grassland is frequently ploughed. Land which has been under grass for more than seven years is usually called 'permanent pasture'.

This term is often also applied to some types of grassland which have never been ploughed, while others, such as steep valley sides with acid soils, are often referred to as 'rough grazing'. To avoid confusion, 'semi-natural grassland' will be used here for those which have probably never been ploughed, or not for a long time. Examples are flood meadows, the calcareous grasslands of steep valley sides in limestone country, and acid grasslands. Their importance to the naturalist is that they have not originated from sown seeds and usually contain a relatively rich flora and fauna.

Grasses are thought to improve the structure of agricultural soils and it is still common practice to put fields down to grass for short periods as part of a rotation system. Such grasslands are known as 'leys'. Clovers are often included in the seed mixture, for they improve the fodder value of the pasture and also the nitrogen content of the soil. The nodules on the roots of clover contain bacteria which are capable of 'fixing' nitrogen from the atmosphere.

Over most of Europe, the grasslands have been derived from forest and would return to it if the grazing animals were permanently removed. Indeed, in many localities where poorer grazing has been abandoned, the land has reverted to scrub. In northern and southern Poland and lower-lying parts of western Czechoslovakia, however, the flora of certain small areas resembles the extensive steppes of Hungary and the Ukraine where, because the climate is so hot and dry in summer, forest did not exist even in primeval times. These Polish and Czechoslovakian sites are particularly well developed on steep, south-facing slopes where the underlying rock is limestone (calcium carbonate) or gypsum (calcium sulphate); they are therefore exceptionally dry.

How grasses survive grazing

Flowering plants are separated into two groups: the monocotyledons, to which the grasses belong, and the dicotyledons, which include many of the plants with broad leaves and brightly coloured flowers. The names refer to the number of cotyledons, or seed-leaves, carried by the embryonic plant, but, from the ecological aspect, other characteristics are of more significance.

The leaves and flowers of all flowering plants originate in buds which are produced above or below the soil surface. Many herbaceous dicotyledons produce buds at heights varying from a few centimetres to two metres or more above the ground and when these are grazed the buds, as well as the mature stems and leaves, are lost. If a plant of this type is to survive, new buds must be produced on some part of the plant that remains and these must grow into new, leaf-bearing shoots. Until this happens, food production by photosynthesis ceases, as this process is carried out mostly by the leaves.

The buds of grasses are rarely damaged by grazing animals because they are produced close to the soil surface. Another important distinction between dicotyledons and grasses is the manner in which their leaves grow. Leaves are composed of large numbers of microscopic units, the cells, and in dicotyledons their growth is essentially a two-stage process, the first of which is cell production and the second cell expansion. In the grasses, on the other hand, a growing point at the base of the leaf continues to produce new cells long after those on the upper part have expanded and matured. Thus soon after the upper part of a grass leaf is eaten (or mown), new photosynthetic tissue is formed from the base. These differences give the grasses a competitive advantage over many of the dicotyledons under heavy grazing pressure.

Management of grasslands

The management of grasslands varies enormously. In parts of southern Germany, for example, grass is cut for hay and it is unusual to see cattle and sheep out in the fields. In contrast, the hill pastures of Wales were, at least until the early part of the 20th century, grazed by sheep all the year round: in winter, the wethers (castrated males) were left on the hills, to be joined by the ewes and lambs in late spring. More recently, however, the demand for mutton has greatly declined so the Welsh hills are grazed only in summer.

A common form of management is to cut the grassland for hay in early summer and then allow sheep and cattle to graze the aftermath. Sometimes, however, pastures are grazed in winter. This has the effect of breaking up the leaf litter, some of which is eaten, which in turn promotes the growth of young shoots in the spring. In some areas, rough grazing is burned in late winter or early spring to destroy the litter and encourage an 'early bite' for the animals. This type of management has a profound effect on plant and animal populations. Domestic animals are selective and overgrazing can so inhibit the growth of palatable plants that patches of ground may be exposed and erosion may then set in. On the other hand, if the pasture is undergrazed, the less palatable species are scarcely eaten and, since they often grow vigorously, are likely to spread at the expense of the others. In hay fields the proportions of different species can be affected even by the time of cutting. Burning kills a large proportion of the fauna and winter grazing reduces the numbers of invertebrates because the habitat in which they live, the litter layer, is destroyed. All types of management are likely to affect the frequency of inflorescences and since many insects, such as butterflies and bees, feed on nectar and pollen, they too will be affected.

Removal of animals or hay from grassland involves a loss of nutrients, and burning of pasture liberates to the atmosphere a large proportion of the nitrogen held in the aerial parts of the plants. Other nutrients, such as phosphate and potash, are deposited with the ash. On riverside meadows losses are made good naturally when silt is deposited during flooding, but on other grassland types there are only the small amounts of nutrients in the rain and those released during the slow weathering of soil particles to offset what is lost. Unless the soil nutrient content is maintained the proportions of different plants will change and nowadays it is becoming increasingly common to fertilize grasslands.

Grasslands on chalk and limestone

Chalk and limestone are different forms of the same mineral, calcium carbonate. Grasslands overlying these rocks are known as calcareous grasslands. Because chalk and limestone are permeable to water, these grasslands are relatively dry, especially on warm, south-facing slopes.

The softest of these rocks is chalk, which forms low, rounded hills in England (from Yorkshire and Lincolnshire to east Dorset and thence east

A dry chalk valley invaded by scrub. Its steep sides are grazed by sheep whose tracks are visible on the right of the valley; Chiltern Hills.

to Kent and Sussex), in northern Denmark, and in France (from the Pas de Calais north of Paris to the middle Loire and then back south of Paris to Champagne). In the Limburg area of northwest Germany, as in the Normandy area, much of the chalk is covered by loess.

Extensive outcrops of the somewhat harder oolitic limestone occur in England (from Yorkshire to Dorset), in eastern and central France (notably the Langres plateau), in Bavaria and in southern Poland. Carboniferous limestone, the oldest and hardest form, is exposed over large areas in the English Pennines and in western Ireland. On the chalk and oolitic limestone the grasslands are usually on steep valley-sides which cannot be ploughed, but on the carboniferous limestone they are widespread, often in a severely eroded state, on gentle slopes as well.

Calcareous rough grazings are rich in plants and as many as 40 or more species may occur in a square metre. The proportions of the different species vary enormously from one locality to another, however, depending on such factors as management, aspect, climate and geographical location.

In Britain, many of the plants of calcareous grasslands are almost restricted to this habitat, but others grow in a variety of vegetation types. Examples of the former are the pasqueflower *Pulsatilla vulgaris*, traveller's-joy *Clematis vitalba*, horseshoe vetch *Hippocrepis comosa*, common rock-rose *Helianthemum chamaecistus* and downy oat-grass *Helictotrichon pubescens*, while the sheep's-fescue *Festuca ovina* grows equally well on either calcareous or acid grassland. It is likely, however, that some of the plants common on calcareous grassland grow there because they can tolerate the dry conditions and are unable to compete with other species on the deeper, moister loams.

Certainly, many of the plants of limestone appear to be well adapted to dry conditions. Measurements of water in the chalk have shown that, at a depth of only 20 cm, there is sufficient throughout the summer to prevent plants wilting, and the roots of many species penetrate to considerably greater depths: for example, those of the salad burnet *Sanguisorba minor* reach to 50 cm, of the common rock-rose to 75 cm, and of the horseshoe vetch to as much as 90 cm. Beneath the thin soil, the upper layers of limestone have extensive fissures and it is by exploiting these that the roots are able to penetrate to such depths.

Other species are able to survive dry periods by lowering the rate of transpiration (water loss) from their leaves. Thus, during dry periods, the leaves of some grasses fold, so that the upper sides are brought together, while those of others roll into cylinders. In both cases, the area of the transpiring surfaces is greatly reduced. The narrow leaves of sheep's-fescue are permanently folded.

Most of the plants of calcareous grasslands are perennials. The few annuals, which are generally shallow-rooted, evade drought conditions by completing flowering and seed-production before mid-summer.

Soils containing calcium carbonate are mildly alkaline and, in these conditions, iron and phosphorus form chemical compounds of low solubility. Iron is a component of the green pigment chlorophyll and, if plants are unable to absorb enough, they become yellow, or chlorotic. Even some of the species characteristic of calcareous soils, which are generally adapted to the low levels of available iron, may become chlorotic locally. Iron and phosphorus deficiencies are other factors which, like the dryness of the soil, probably prevent many plants of neutral or mildly acid soils from growing on limestone areas.

On many of the lower-lying limestones, tor-grass *Brachypodium pinnatum* seems to be spreading rapidly and this is often considered a result of undergrazing. Indeed, many areas of calcareous grassland are not now grazed at all by sheep and, following the spread of myxomatosis, relatively little by rabbits. Rather few plant species grow where the tor-grass is abundant and so its spread is likely to lead to an impoverishment of the grasslands concerned. Another result of undergrazing is that shrubs are able to invade the grassland. Where these become dense, the herbaceous plants die because they no longer receive enough light.

Acid grasslands

Rain contains small quantities of nitric and, particularly near industrial centres, sulphuric acids. As the water percolates through limestone soils, fragments of calcium carbonate remaining in the soil neutralize these acids and the soils stay on the alkaline side of neutrality. In the absence of calcium carbonate, however, the acids are not neutralized and, as a result, essential metals such as calcium and potassium are leached away from the rooting layers. Thus the soil becomes progressively more acid until, at a certain stage, aluminium, which is soluble only in acid conditions, dissolves in the soil moisture. This metal is toxic to many plants, but the plants which are characteristic of acid grasslands are, to varying extents, tolerant of aluminium. The higher the rainfall, the greater the degree of leaching, so the extent of acid grasslands increases towards the Atlantic coasts and also with altitude. Such grasslands also occur on readily leached sandy soils in areas of relatively low rainfall.

Even from a considerable distance acid grassland on a hillside appears to be a mosaic of different vegetation types. Ribbons of dark green vegetation – the flushes – can be seen where water collects and drains downhill and, between these, patches of browner vegetation grow on drier hummocks of glacial till or on shallow soils where the rock is close to the surface. The plant species growing in the flushes will differ from those on the drier areas, as a result of differences both in the amount of water available and in the mineral content of the soil. Sheep's-fescue and common bent *Agrostis tenuis* are the most common grasses on the drier areas, with sweet vernal-grass *Anthoxanthum odoratum* and cock's-foot *Dactylis glomerata* on the somewhat richer

Selective grazing by sheep on acid grassland has left large patches of the unpalatable heath rush; Yorkshire.

parts. Along the flushes, purple moor-grass *Molinia caerulea* is usually abundant, especially where the gradient is shallow and peat tends to accumulate, since this species can tolerate waterlogged soil. Where the gradient is steep, a few plants of great wood-rush *Luzula sylvatica* sometimes survive, indicating that the hillside was once covered with forest. On the gentler slopes between the flushes, mat-grass *Nardus stricta* may be abundant, especially at higher altitudes.

'Neutral' grasslands

Across the lowlands of Europe, there are many different types of grassland which are moist throughout the year, even flooded for short periods, and only slightly acid. The great variation in their flora depends on such factors as wetness, the nature of the soil, geographical location and, above all, the way in which they are managed. One of the most common types occurs where the water-table is permanently high. It is characterized by purple moor-grass and develops where fens are regularly mown or where wet woodlands have been felled and the land subsequently grazed. When mown, this type of grassland rarely gives more than one cut a year.

Land situated alongside natural rivers is subject to flooding in winter and early spring and is, in effect, fertilized by the silt deposited when the water recedes. Here grasslands with a rich variety of plant species occur because they are too wet to plough, and so have been left to themselves for many centuries. In Britain, meadow foxtail *Alopecurus pratensis* and sweet vernal-grass are among the most characteristic grasses and these are accompanied by many herbs typical of moist habitats. In addition, the irrigation of many flood meadows has, in some measure, been brought under control by the construction of weirs, water channels and embankments. In these water meadows, efforts are made to suppress many of the herbs to the advantage of the grasses, and this treatment seems to favour the large fescues. Both tall fescue *Festuca arundinacea* and giant fescue *F. gigantea* are particularly common. In Poland, as a result of river draining schemes, the majority of lowland meadows are no longer flooded. They are still very productive, however, because of regular treatment with artificial fertilizers. Two or even three cuts can be taken every year and, with this type of management, the false oat-grass *Arrhenatherum elatius* becomes characteristic. The composition changes markedly when grazing is substituted for hay cropping.

Cereals do not yield well in a cool, damp climate and the amount of land under grass has increased steadily since the middle of the 19th century. In northern Normandy, including the Cotentin peninsula, grassland now covers as much as 60 per cent of the land. A similar situation exists on the older reclaimed land in the district of Holland and in the south of the Netherlands. In coastal areas of the extreme north of the Netherlands, nearly 80 per cent of the land is under grass but this is short-term ley. Much of Denmark is permanent grass but it is ploughed and re-sown from time to time.

Development of scrub

Shrubs are often present on semi-natural grasslands and frequently in far greater numbers than a casual glance might suggest. Many are only a few centimetres high and heavily grazed, but, nonetheless, they can be quite old. If grazing pressure is reduced, these may be able gradually to increase in size since it is likely that they will then retain a greater proportion of each year's growth. There was a sharp reduction in grazing pressure over much of Britain in the 1950s as a result of the spread of myxomatosis among rabbits and, together with the decline of sheep- and cattle-rearing over the last decade or so, this has resulted in the spread of scrub over many areas of grassland. Paradoxically, in some parts of Europe improvements in agricultural technology have also contributed to the development of scrub: steep slopes that were once tilled by hand cannot be worked with modern agricultural machinery and, since it is not possible to make a living by the old methods in competition with the new ones used on the richer soils, the hill slopes have been abandoned. Scrub has increased greatly in France over the last century, but most of it is in the south. There is a certain amount on the oolitic limestone of the northeast – on the Langres plateau, for example – but very little on the chalklands of Champagne and Normandy, which are either intensively grazed or ploughed.

The scrub that develops on limestone is often particularly rich because, as well as those shrubs which occupy a wide range of soils, such as hawthorn *Crataegus monogyna*, blackthorn *Prunus spinosa*, elder *Sambucus nigra* and hazel *Corylus avellana*, certain calciphilous (lime-loving) species may be present. These include dogwood *Cornus sanguinea*, buckthorn *Rhamnus catharticus*, privet *Ligustrum vulgare* and spindle *Euonymus europaeus*. Such rich associations are common in southern Britain, but on the Yorkshire Wolds scrub consists almost entirely of hawthorn, with some elder and blackthorn, although the area is within the range of several calciphilous shrubs. These wolds have, however, been farmed relatively intensively for many centuries; even at the time of the Norman invasion there was scarcely any woodland and probably little scrub. Thus the only shrub available on a large scale to provide seed for colonization is the hawthorn of the hedges. Calciphilous species are absent from most of north England where the carboniferous limestone is exposed; presumably this is a result of the high altitude, for they do occur in the valleys at the southern end of the Pennines.

As the shrubs spread, so the grassland herbs decline. On one area of chalk, the average number of grassland species in each sample patch was 31, while in a neighbouring area of scrub with closed canopy it was only one.

On acid grasslands, particularly on dry sands and gravels, gorse *Ulex europaeus* and broom *Sarothamnus scoparius* are likely to become established. On wet, neutral grasslands, reversion to scrub alder *Alnus glutinosa* and willows *Salix* occurs when grazing ceases, though it is relatively rare for such areas to be abandoned altogether, because the soils are usually fertile and likely to be ploughed if they are no longer required for grazing or hay.

Many of the shrubs mentioned are thorny. The permanent thorns are produced during the second year of growth and are lateral shoots which cease to elongate after one year. The growing point at the apex of the branch dies and the tip becomes pointed and hardened. (Short thorns are sometimes produced on the normal shoots of hawthorn during the first year.) The abundance of thorns makes thickets difficult to penetrate, and other plants, such as tree seedlings and saplings, are accorded added protection from grazing animals. As these saplings grow, the scrub gradually takes on the appearance of woodland.

Nutrient cycling in grassland

Especially where grassland is not intensively grazed or managed for hay, much litter is deposited at the end of the growing season. This is attacked by micro-organisms and, as in woodlands, various animals assist in the decomposition process. A proportion of the material is consumed and deposited as faeces, and some of the rest is broken up into small fragments. The faeces and fragments are decomposed relatively quickly by bacteria and fungi.

The most abundant animals in grassland litter are usually springtails (Collembola) and mites (Acari), though many of these may feed on the fungi rather than the litter itself. Because they are so small, their impact is not great. Woodlice (Isopoda) feed directly on the grass litter, but the moisture regime has a major effect on their distribution. Where the litter is deep and the lower layers therefore damp, large numbers of *Trichoniscus pygmaeus* and *T. pusillus* may occur, but in the litter of calcareous grasslands species more tolerant of drier conditions may be present, such as *Philoscia muscorum*, *Porcellio scaber* and *Cylisticus convexus*. In general, the woodlouse fauna of Britain is rather limited, apparently because of the low summer temperatures, and more species are likely to be encountered on the mainland of Europe.

The moisture regime also has a marked effect on the distribution of molluscs, but snails in particular have an additional high requirement for calcium which, in the form of calcium carbonate, is the chief component of their shells. The calcareous grasslands, which are those most rich in calcium, are also dry and so the animals that live there must be among the more drought-resistant. Typical inhabitants of short, calcareous turf are the striped snail *Cernuella virgata* and heath snail *Helicella itala*, two species which are found only in habitats rich in lime. Their largely white shells reflect much of the sun's radiation and they often climb the stems of grasses in hot weather, apparently to get away from the warm soil.

Most snails lay their eggs in summer or have no specific breeding season, but the striped snail and wrinkled snail *Candidula intersecta* are among those, characteristic of dry habitats, that breed in autumn and winter. Unlike many other snails, they seem resistant to cold weather and do not hibernate. In this way, the young avoid the difficult conditions of summer while they are in their most vulnerable state.

Other species frequent in calcareous grassland are the grove snail *Cepaea nemoralis*, beautiful grass snail *Vallonia pulchella*, Kentish snail *Monacha cantiana* and hairy snail *Trichia hispida*. These, however, are also often found in other habitats as well. In damp meadows, slippery snails *Cochlicopa lubrica* and crystal snails *Vitrea crystallina* may be common, but these too occur in a variety of places. In general, the moisture regime and supply of calcium carbonate appear to have far greater effects on the distribution of molluscs than the type of vegetation.

Earthworms (Lumbricidae) can be more abundant in grassland soils than in any other lowland habitat, but, since the supply of litter is usually small, most ingest the soil, and absorb bacteria and other micro-organisms from it.

Plant-eating soil-dwellers

The plant-eating larvae of several groups of insects feed on the underground parts of grasses and can have a considerable effect on their growth. Among the most damaging are those of various species of crane-flies (Tipulidae) and click-beetles (Elateridae), known respectively as 'leatherjackets' and 'wireworms'. When grassland is ploughed and sown with cereals or, for example, potatoes, the larvae turn their attention to the crop. The numbers of these

Heath snails on salad burnet. Snails often climb the stems of grasses in hot weather, apparently to get away from the warm soil.

insects are relatively low in soil that is ploughed regularly, but, since wireworms may take up to three years to mature, crops can be seriously damaged for that length of time after grassland has been turned over.

Where there is adjacent woodland, considerable damage can be done to roots by the larvae of a cockchafer or maybug *Melolontha melolontha*. These also take several years to develop and will attack the underground parts of several crops, particularly cereals, beet and potatoes; the adult beetles feed mostly on the leaves of various trees. The caterpillars of some moths, notably the flounced rustic *Luperina testacea* and a grass-moth *Chrysoteuchia culmella*, also feed on grass roots.

Plant-eating insects feeding above ground

The young stages and adults of many insects feed on the aerial parts of plants. Among those most closely associated with grassland are butterflies and moths (Lepidoptera) and grasshoppers and bush-crickets (Orthoptera).

The caterpillars of most butterflies are exclusively vegetarian, but, because each species is dependent on only a few food-plants (or, in some cases, just one), not many different butterflies are seen over short-term leys or even certain types of permanent pasture. On the other hand, many species can often be found on semi-natural grassland with a rich variety of plant species, especially in the warmer parts of the European mainland. The adults there

The caterpillar of a large blue butterfly rearing up for adoption by an ant. The caterpillar will be carried to the ants' nest where it will feed on ant larvae.

also have a wide choice of flowers from which to take nectar. The butterfly families most widely represented on European grasslands are the Lycaenidae (hairstreaks, coppers and blues) and the Pieridae (whites and yellows). Many of the blues feed on plants belonging to the pea family (Leguminosae), and the coppers on the dock family (Polygonaceae). The whites and the orange-tip *Anthocharis cardamines* feed mostly on the cabbage family (Cruciferae), while the clouded yellows *Colias* again favour the Leguminosae.

The life histories of some of the Lycaenidae show unexpected features. When small, the caterpillars of certain species, notably of the long-tailed blue *Lampides boeticus* and small blue *Cupido minimus*, have cannibalistic tendencies. More remarkable, those of the long-tailed blue and certain others, such as the chalk-hill blue *Lysandra coridon* and the large blue *Maculinea arion*, also associate with ants. From a special gland about half-way along the body, these caterpillars secrete a sweet substance which is avidly eaten by ants. The ants often carry a caterpillar to the neighbourhood of their nest, depositing it on the correct food-plant, while certain ants, especially *Myrmica scabrinodis*, even carry the caterpillars of the large blue into their nests, where they allow them to feed on their own larvae.

Many moths live on grasslands, but few are as brightly coloured as the six-spot burnet *Zygaena filipendulae* and its relatives. Like those of numerous other species, the caterpillar weaves a silken cocoon about itself before it pupates and this presumably has a protective function. The pale yellow cocoon of the six-spot burnet is often conspicuous, attached high up on a grass stem. Since its caterpillars tend to be social, several cocoons can usually be found in a small area.

The bodies of grasshoppers (Acrididae), crickets (Gryllidae) and bush-crickets (Tettigoniidae) are green, brown, or a mixture of the two, and the disposition of the colours is such as to break up the outline of the body and thus provide camouflage. Having, in effect, hidden themselves from predators, they advertise their presence to each other by stridulating persistently: this involves the production of sounds by rubbing the forewings together (crickets and bush-crickets) or hindleg on forewing (grasshoppers). Usually only the males stridulate, and experimental work with recordings has shown that the females move towards the source of the sound. Crickets and bush-crickets have a sensory organ by which the sound is perceived, the tympanum, situated on the foreleg, whereas grasshoppers have it at the base of the abdomen. The rate of stridulation increases with temperature (indeed, for some species, it has even been claimed that the temperature can thus be estimated), but the character of the 'song' remains constant and it is not difficult to identify many of these insects by the sounds they produce.

Towards the end of the summer, grasshoppers, bush-crickets and some crickets lay their eggs in the soil or at the bases of grass tufts. The eggs begin to hatch in spring, and the first adults appear in June. After a brief, worm-like, larval stage, the young or nymphs resemble small, wingless adults; subsequently, depending on the species, they moult four to ten or more times as they grow and their wings, at first rudimentary, develop gradually at each stage. In general, adult bush-crickets and grasshoppers do not survive the winter in Europe. Grasshoppers feed mostly on grasses, but crickets and bush-crickets eat many small insects as well.

The bugs (Hemiptera) are another order of insects widely represented in grasslands; they are divided into two suborders, the Homoptera and the Heteroptera. Well-known among the former are the spittle-bugs or frog-hoppers (Cercopidae), the nymphs of which enclose themselves in foam popularly known as 'cuckoo spit'. Several of the Heteroptera may be especially common in moist grasslands. Some, such as the capsid *Notostira elongata*, overwinter as adults, but the meadow plant-bug *Leptopterna dolobrata* and many others die and only their eggs survive the cold weather. The capsid mentioned is relatively successful in grasslands that are frequently cut, possibly because it produces two generations during the summer.

The brightly coloured flowers of grasslands attract many different insects that feed on pollen and nectar, but there are also some that eat the petals. Earwigs (Dermaptera) consume the actual tissue, while thunderflies or thrips (Thysanoptera) suck the sap and leave the petals marked or deformed. Soldier-beetles *Rhagonycha fulva* are another familiar sight on flowers, but they are predatory on other visiting insects.

Predatory insects and spiders

The damsel-bugs (Nabidae) are a family of predatory insects characteristic of, and almost limited to, grasslands. They simply leap on other insects and, holding them with their forelegs, suck the liquids from their victims' bodies through a beak-like rostrum. The technique is sufficiently effective for them to be able to feed on other animals of about their own size: the field damsel-bug *Nabis ferus*, for example, often preys on the meadow plant-bug.

Various spiders also live in grasslands. The hammock-spider *Lynyphia triangularis* commonly slings its hammock-web from the stems of grasses. The purse-web spider *Atypus affinis*, on the other hand, spends most of its life in a buried silk tube which extends on to the surface shaped like a finger of a glove. When an insect settles there, the spider enters this part of the tube and transfixes its victim from below, cuts a hole through which the prey can be dragged, and finally repairs the hole.

Many spiders catch their prey without the aid of a web. The crab-spider *Xysticus cristatus*, for example, waits for an insect to come within striking distance, then grabs it with well-developed forelegs. Another spider *Evarcha falcata* also has powerful forelegs with which it seizes its prey, often in mid-air: this species waits on a vertical surface, to which it attaches itself by a silken thread, jumps at a passing insect and then, with the aid of the thread, swings back.

Amphibians and reptiles

Dry grassland, especially where it tends to give way to bare patches of earth, gravel or rock, is a habitat for some of northern Europe's few reptiles, such as the wall lizard *Lacerta muralis*, green lizard *L. viridis* and viviparous lizard *L. vivipara*. The last may be found in quite damp areas, also the home of the common frog *Rana temporaria* and common toad *Bufo bufo* outside the breeding season.

Herbivorous mammals

The available evidence indicates that, in general, the effects of plant-feeding insects on grassland production are small: treatment with a powerful insecticide, for instance, results in little change in the amount of herbage. On the other hand, certain mammals, such as the rabbit, consume considerable quantities of vegetable matter and thus have a marked effect on the rate of nutrient circulation. In Britain, after the introduction of myxomatosis in 1953, many grasslands, formerly rabbit-infested, became far more lush and the density of flowers much greater. Within a few years, tussocky grasses had developed to such an extent that the associated herbs were often suppressed.

Normally voles consume relatively small amounts of vegetation in British grasslands, but occasionally the numbers of field voles *Microtus agrestis* become so great that considerable damage is done both above and, as a result of their burrowing, below the ground. On the European mainland, such

Predator and prey. LEFT *A damsel-bug sucks the juices from a captured fly using its beak-like rostrum.* RIGHT *A bumblebee is caught by a crab-spider lurking amongst bramble flowers.*

'plagues' are relatively frequent; there, the species usually concerned is the common vole *M. arvalis.*

The number of voles builds up over a period of several years and then crashes dramatically. Much research on these fluctuations seems to have eliminated the more obvious possible explanations, such as food shortage, weather, disease, parasites and predation. It seems feasible that complicated genetic selection may be the cause, less aggressive types increasing with the rest of the population during the phase of increase, but then being eliminated by more aggressive strains when overcrowding occurs.

The most effective grazers are usually domesticated sheep and cattle, which consume a large proportion of the annual production of plant material in grasslands. Since only about ten per cent of their intake is retained within their bodies, they promote a rapid circulation of nutrients. The discharged material is, however, locally deposited and, as domesticated animals will not graze near their own droppings, the carrying capacity of an area would decrease rapidly were it not for the activities of soil-inhabiting invertebrates, such as earthworms, dor-beetles *Geotrupes* and dung-beetles *Aphodius.* Earthworms feed directly on the droppings and, since they mostly discharge their casts below the surface, they play an important part in mixing the dung with the mineral soil. The beetles excavate burrows, carry fragments of dung down into them and lay their eggs close by; the dung then serves as food for the larvae.

Parasites of mammals

Domesticated animals act as hosts to numerous parasitic flies. The Tabanidae, for example, which include the cleg *Haematopota pluvialis*, the horse-fly *Chrysops caecutiens* and the stout *Tabanus sudeticus*, have piercing mouthparts with which they can penetrate the skin of the host and extract blood from peripheral blood-vessels. Only the females feed in this way, blood apparently being necessary for the development of their eggs; the males feed on nectar. The larvae also have piercing mouthparts and suck the body fluids of a wide variety of invertebrates. Two more blood-sucking flies of other families are the stable-fly *Stomoxys calcitrans*, the larvae of which develop in straw fouled by cattle, and the forest-fly *Hippobosca equina,* which actually lives on the bodies of horses and cattle.

Adults of the warble-fly *Hypoderma bovis* have only vestigial mouthparts and do not feed at all, but the larvae cause considerable damage to the hides of cattle. The warble-fly attaches an egg to a hair on the leg of a beast, and the newly-hatched larva enters the skin at the base of the hair. Subsequently, it makes its way through the body of the host to a position just below the skin of the back and the fluids which accumulate around the parasite cause a lump or 'warble' to develop there. After about two months, the parasite emerges through the skin and drops to the ground, where it pupates.

Greenbottles *Lucilia* sometimes lay their eggs in open wounds on domesticated animals and the larvae feed on the flesh.

Predatory mammals

Although some of the European carnivores, such as the red fox *Vulpes vulpes* and stoat *Mustela erminea*, are often seen in grasslands, they also occur in afforested land and, since this is the natural vegetation type over most of Europe, they must be considered basically woodland species. On the other hand, the shrews (Soricidae) although frequent in woodland are commonest in tussocky grasslands, scrub and hedgerows. Thus, grasslands with good cover are probably their main natural habitat.

The water shrew *Neomys fodiens*, common shrew *Sorex araneus* and pygmy shrew *S. minutus* are widespread in lowland Europe, though only the last occurs in Ireland. The common white-toothed shrew *Crocidura russula* and bicoloured shrew *C. leucodon* are absent from Britain (though the former is found in the Channel Islands), but are well-distributed on the European mainland. The lesser white-toothed shrew *C. suaveolens* appears to avoid the northern lowlands of Europe but is the only shrew on the Isles of Scilly.

Shrews are extremely active and feed voraciously, largely on invertebrates. Their sight is very poor, but their senses of smell and hearing are highly developed. Most of them favour dry grasslands, but the common shrew also occurs often in marshes. Breeding usually begins early in May, but few appear to survive for more than a single winter; none of the species hibernates.

Birds of grassland and scrub

Comparatively few birds breed on open grassland, but they range in size from the pipits *Anthus*, skylark *Alauda arvensis* and crested lark *Galerida cristata* to gamebirds (Phasianidae), lapwing *Vanellus vanellus*, stone-curlew *Burhinus oedicnemus* and little bustard *Tetrax tetrax* up to the now rare great bustard *Otis tarda*, one of Europe's largest landbirds. Pipits' nests may be well hidden in a tussock, while stone-curlews and lapwings prefer almost bare areas, on which they may be joined locally by ringed plovers *Charadrius hiaticula*, usually found on beaches and river gravel. The larger species tend to display on the ground and in the open, while the larks and pipits sing in the air.

Scattered clumps of shrubs, covering only a small proportion of the grassland, may bring such birds as the tree pipit *Anthus trivialis*, dunnock *Prunella modularis*, whitethroat *Sylvia communis*, grasshopper warbler *Locustella naevia*, linnet *Carduelis cannabina*, yellowhammer *Emberiza citrinella* and reed bunting *E. schoeniclus* into the community. The presence of the meadow pipit *Anthus pratensis* and dunnock, and perches from which their movements can be watched, are likely to attract the cuckoo *Cuculus canorus*, which often lays its eggs in their nests.

A few trees will make the area suitable for several more species. If partially dead and hollow, they may well attract hole-nesters, including the green woodpecker *Picus viridis* and wryneck *Jynx torquilla*. The green woodpecker often feeds at nests of ants in open grassland and these insects are also the food of the wryneck. The numbers of wrynecks in western Europe have been decreasing steadily for more than a century – as have those of the red-backed shrike *Lanius collurio*, which also occurs in open country with clumps of thorny bushes. The reasons for these declines are obscure, but it has been suggested that the gradual climatic deterioration towards cooler, wetter summers has affected the populations of the insects on which these birds feed. More recently, there has also been a sharp drop in numbers of woodlarks *Lullula arborea* in western Europe, accelerated by the severe winters of 1961/62 and 1962/63, and this has been linked with the spread of myxomatosis: although they nest in longer grass, woodlarks usually feed on short swards and, with fewer rabbits, the vegetation has tended to grow taller.

As the proportion of scrub increases, so does the density and variety of birds. Well-developed scrub, covering 80 per cent or more of the land, may support well over twice the number of species and individuals that grassland with scattered bushes does, despite the fact that most of the birds characteristic of the open habitat are no longer present. Several buntings appear at the scrub stage, and willow warblers *Phylloscopus trochilus* may be abundant when it becomes dense. Bush-nesting warblers and the turtle dove *Streptopelia turtur*, another summer visitor, are also common; the nightingale *Luscinia megarhynchos* and its northeastern counterpart, the thrush nightingale *L. luscinia*, sing from thickets with low cover for nesting.

The small breeding birds of open grassland feed mainly on invertebrates, as do stone-curlews and lapwings. In general, nestlings of mainly seed-eating birds, such as larks, finches and buntings, are fed on insects.

Wet grasslands and flood meadows support a greater variety of breeding birds than dry ones. The lapwings may be joined by snipe *Gallinago gallinago*, curlews *Numenius arquata* and redshanks *Tringa totanus* (both scarce in France), as well as black-tailed godwits *Limosa limosa* and ruffs *Philomachus pugnax* (two species that have recently returned to breed in Britain). Several ducks may also nest in long grass some way from water. Reed buntings and grasshopper warblers are typical small birds of this habitat, though both may be found in dry areas as well.

Skylarks and lapwings congregate in flocks in winter. They continue to feed on grasslands, especially lowland pastures, and, in Britain, Ireland and adjacent parts of the Continent, are joined by golden plovers *Pluvialis apricaria*, starlings *Sturnus vulgaris* and many thrushes, including fieldfares *Turdus pilaris* and redwings *T. iliacus* from farther north and east. In winter, all of these species, except the fieldfare, are found only west of Poland, where the climate is relatively mild. Indeed, outside Britain and Ireland, the golden plover is limited to the coastal areas of France and the Low Countries.

Raptors

Although most of the falcons of northern and central Europe frequently breed in trees, using the old nests of other large birds, they all hunt over open country. Man's destruction of forests must have benefited these birds and it is interesting that the breeding ranges of some species, notably the red-footed falcon *Falco vespertinus* and lesser kestrel *F. naumanni*, do not extend beyond where the natural grasslands occur. Both these and the hobby *F. subbuteo* are migratory, wintering in Africa. Sizeable insects, especially grasshoppers (Orthoptera), dragonflies (Odonata) and beetles (Coleoptera), form a large proportion of their diet, though the hobby takes many small birds as well, particularly in the breeding season, and the other two also feed their young to some extent on lizards and voles. The kestrel *F. tinnunculus* is only a partial migrant, those that breed in northern and eastern Europe moving into central and western parts in winter; it is unusual among European falcons in that it feeds largely on voles and mice, especially *Microtus* and *Apodemus*.

The kites prey upon a wide variety of mammals, reptiles, amphibians and young birds, and are also partial to carrion, including dead fish. On the Continent, the black kite *Milvus migrans* is the commoner species, though it is only a rare vagrant to Britain; it sometimes gathers in flocks at carrion and also tends to be sociable during the breeding season, when several pairs may nest close together. Until the end of the 18th century, the red kite *Milvus milvus* was numerous in Britain, even invading towns to feed among the rubbish littering the streets, but in the following hundred years it was so

A newly-hatched cuckoo in a meadow pipit's nest tips out the last of its host's eggs.

persecuted, in the interests of game preservation, that it is now confined to upland Wales. On the Continent, it is found mainly in lowlands with scattered woods and cultivated areas, but there too it is becoming rare.

Interest in shooting game followed the development of the double-barrelled shotgun. In British law, game animals are the property of the owner of the land on which they are found and the game laws of the early part of the 19th century made it a criminal offence to trespass in search of them. Keepers were appointed to protect and breed pheasants *Phasianus colchicus* and partridges *Perdix/Alectoris*, and they made it their business to destroy raptors, which it was considered might kill these. Large birds of prey, such as the buzzard *Buteo buteo* and golden eagle *Aquila chrysaetos*, have also been destroyed by hill farmers and shepherds because of the belief that they kill young lambs.

In spite of modern legislation to protect birds of prey in most countries of northern Europe, many are still destroyed, especially when they are nesting and most vulnerable; the commercial value of young raptors to be kept for falconry or prestige is an additional threat in recent years.

ARABLE LAND

Farming trends

Much of Europe, especially in France and western Germany, is still farmed on the open field system: each of the farmers in a village community works a number of strips in each of the surrounding fields. Efforts are being made in both the countries mentioned to consolidate these, but the average farmer's holding is still small. As consolidation proceeds and more people move from rural to urban environments, it is to be expected that agriculture will follow the directions taken in Britain, Denmark and the Netherlands, where it is as advanced as anywhere in the world.

Until about the mid-20th century, most British farmers aimed to be more or less self-sufficient. Income was derived from the sale of animals, dairy products and cash crops, while other crops were grown to feed the sheep and cattle. By feeding animals on root crops, such as turnips and swedes, and grazing them on pasture, the farmer ensured that dung was added to the soil. Additionally, by practising some form of rotation, he saw to it that each field received its share of organic material, none was exhausted of mineral nutrients by the heavy demands of a particular crop, and animal and plant pests were kept under control.

Today, however, more and more farmers are specialists. This has been made possible by the greatly increased use of inorganic fertilizers, the availability on a large scale of herbicides and pesticides, great improvements in the efficiency of farm machinery and changes in rearing methods. Thus, many farmers have been able to specialize in arable and grow the same crop in a field year after year, because they can now control pests and replace mineral nutrients removed by the crop quickly and efficiently. Others have been able to concentrate on meat production by keeping animals indoors and, in some cases, buying the whole of their food requirements; this applies particularly to pig-farming.

Although Denmark specializes in meat and dairy products, less than 50 per cent of the agricultural land there is under grass and about two-thirds of that is rotated. A large proportion of the arable production – cereals, root crops and potatoes – is used to feed the animals. Permanent grasslands are ploughed and resown at intervals of about 20 years (and so are not rich botanically); animals are allowed to graze them only during the summer. Similar methods are used elsewhere in Europe, for example in northeast Germany and the Netherlands.

Modern farming methods have had a profound effect on the countryside and its wild plants and animals. In eastern England, for example, specialization in arable crops has made hedges a liability and thousands of kilometres of them have been removed. Improvements in engineering techniques have enabled farmers to install under-drainage systems, and so many wetland areas have been lost; even the main drains, formerly open, are often piped and filled. The availability of inorganic fertilizers has made it possible to reclaim and farm sections of heathland in Britain, Denmark and Brittany, and the widespread use of herbicides and pesticides has reduced the populations of wild plants and animals on agricultural land. Especially where there are large and intensive livestock units, serious waste-disposal problems have led to considerable pollution of water courses. A remarkable feature of modern farming is that the energy put in by the manufacture and use of tractors and other farm equipment, and by the production of chemicals, is greater, often several times greater, than that taken out as crops, but at present there is no indication of a reversal in procedures.

Farm crops

Many factors influence what crops are grown in a particular area and only a few general comments can be made here. Grassland often predominates in a moist climate, while the emphasis is on cereals where it is dry. Wheat and sugar beet require good soils, whereas cereals, such as rye and oats, and potatoes can be produced on poorer ground. Rye and potatoes are important crops in the Northern Lowlands. Around towns and cities, vegetables are grown to satisfy the demands of the urban populations; in such areas a relatively large input of energy, in the forms of ground preparation, fertilizers and so on, is economically worthwhile.

In Britain, semi-natural grassland is characteristic of steep slopes which cannot be ploughed. On the Continent, vines can often be grown on such sites, provided that the aspect is suitable; steep slopes are often terraced.

Wild plants of arable land

Cultivation has a highly selective effect on plants, and it is not surprising that certain species should occur as weeds throughout lowland Europe. Weeds of arable fields can be regarded as the pioneers of a succession which would ultimately lead to forest, but every year the succession is disturbed by ploughing and reversion to the bare ground stage occurs. Most of the plants are annuals, because they have the best chance of reproducing by seed in the arable management system, but there are a few important exceptions such as common couch *Agropyron repens* and field bindweed *Convolvulus arvensis*, perennials which are able to maintain themselves by vegetative growth, despite the regular disturbance.

The plants most likely to invade an arable field at the bare soil stage are those with efficient dispersal mechanisms. The seeds of poppies *Papaver* are minute and light, and can be carried by the wind, while the fruits and seeds of many other species are transported considerable distances by birds and mammals, either in the alimentary tract or attached to feathers or fur. Large-scale seed-production increases the chances of success, and plants with small seeds often produce great numbers. Single plants of the common poppy *Papaver rhoeas* and prickly sow-thistle *Sonchus asper* may each produce 20,000 or more seeds.

When a field is ploughed, seeds may be buried to a depth of 30–40 cm. If they germinated at that depth, their food reserves would probably be exhausted before the tips of the shoots reached the surface, but the seeds of many arable weeds become dormant when buried, apparently due to the high concentration of carbon dioxide in the soil, and do not emerge until they are again exposed at the surface. Many dormant seeds remain viable for long periods: almost all of those of black nightshade *Solanum nigrum*, for example, are still able to germinate after 40 years.

If weeds are allowed to develop unchecked in a crop, they have a marked effect on the yield. Even the crop plants are competing with each other for some of the essentials of life – light, water and mineral salts – without the additional demands of any weeds. Weeds that reach a height at least similar to that of the crop are likely to affect the supply of light. The wild-oat *Avena fatua*, charlock *Sinapis arvensis* and fat-hen *Chenopodium album* all grow taller than modern, short-stemmed cereals. Furthermore, charlock and fat-hen respectively accumulate relatively large quantities of nitrogen and potassium in their tissues, so the supply of these elements to neighbouring plants is then likely to be insufficient for them to develop their full potential. In areas of low rainfall, all plants are probably competing for water and even low-growing weeds may affect the crop by intercepting significant amounts of rain on their foliage.

Selective weedkillers became available about 1946 and have since proved to be most valuable as a means of controlling weeds. Chemically they resemble the natural growth hormones of plants and, when present in minute quantities, bring about an ordered form of growth; in excessive amounts, however, they can disorganize the growth of weeds to such an extent that they die. These weedkillers are absorbed through the leaves and, since it is the amount taken in that is important, the broad-leaved weeds react more than narrow-leaved cereals, especially since the leaves of the latter are directed upwards so that much of the solution runs off.

Many weeds of arable land, such as fat-hen, common poppy, treacle mustard *Erysimum cheiranthoides*, charlock and common vetch *Vicia sativa*, are sensitive to a wide variety of weedkillers, but several of the bistorts and persicarias *Polygonum* are more tolerant and so relatively difficult to destroy. Nevertheless, surveys in Britain, Germany and Finland, based on results obtained from many different fields, have shown that, even after several years' application of weedkillers, some of the sensitive species,

A massive invasion of arable land by poppies whose seeds are minute and light and therefore easily dispersed by wind.

especially fat-hen and charlock, are still among the most common weeds; the conclusion seems to be that herbicides reduce but do not eliminate weed populations. What is not clear, however, is the extent to which long-term survival of dormant seeds in the soil is responsible for the reappearance of weeds after the use of weedkillers.

During the period of intensive weedkiller usage it has been noted that certain grasses, notably the wild-oat and common couch, have become much commoner in arable fields, especially among cereal crops. These plants are as resistant to most weedkillers as are the cereals; furthermore, if harvested with the crop, they will start a much greater infestation wherever seed from that crop is sown. To obtain a harvest free from wild-oat, many farmers have resorted to removing these weeds by hand.

Invertebrates in arable soil

At the end of the growing season, almost all the above-ground vegetable matter on arable land is carted away or burned. Therefore, many of the animals that feed on surface layers of organic litter in woods and grasslands, such as mites (Acari), springtails (Collembola) and woodlice (Isopoda), are absent. Certain practices, however, provide an alternative source of winter food and enable some invertebrates to survive in relatively large numbers. Slugs (Mollusca) and millipedes (Diplopoda) feed mostly on dead organic material and so the application of manure in autumn provides them with a satisfactory source of winter food; in the spring, at least some species of both will attack newly-emerged seedlings and such autumn-sown crops as winter wheat. The whole root-systems of many crops are left in the soil at harvest time and these help to maintain the food supply for other animals, such as earthworms (Lumbricidae).

The absence of shelter above ground in winter means that insects which feed on crops must either move away or survive in the soil. Some of the latter group – for example, the pea weevil *Sitona lineatus* – pass the winter as adults in a dormant condition, while others, such as the beet-fly *Pegomyia hyoscyami*, overwinter in the pupal stage.

Insects of arable crops

Few crops are likely to be completely defoliated by insects, but the larvae of the Colorado beetle *Leptinotarsa decemlineata* can achieve this in potato fields. The species was accidentally introduced into France from America in about 1922 and has since become firmly established in western Europe; so far, efforts to prevent it spreading into Britain have been successful. Another insect that feeds on potatoes is the larva of the death's-head hawk-moth *Acherontia atropos*, but this species is indigenous to Europe; it switched from its normal food-plants, primarily members of the nightshade family (Solanaceae), such as bittersweet *Solanum dulcamara* and black nightshade *S. nigrum*, after the potato, which belongs to the same family, was introduced into Europe. This large moth, which overwinters as a pupa in the soil, is widely distributed in central and southern Europe, but seems unable to survive the northern winter; migrants appear in southern England usually in June.

Cereals are prone to damage by many different insect species. Two which commonly attack wheat are the wheat midge *Contarinia tritici* and the corn sawfly *Cephus pygmaeus*. Clouds of wheat midges can sometimes be seen drifting across wheat fields and, in bad years, as much as ten per cent of the grain may be damaged by their larvae; the eggs are laid in the florets of the wheat, and the larvae feed on the developing grain for about three weeks. The eggs of the corn sawfly are inserted into the stem, just below the ear, and the larvae subsequently bore their way down to the base. Both these insects spend the winter in diapause (a state similar to hibernation in which metabolic activity is greatly reduced), close to the soil surface. The midge larva encloses itself in a cocoon just below ground and the sawfly seals itself into a short section of the base of the wheat stem, having first felled the upper part by making a neat cut just above soil level.

Insect pests of crop plants may be separated into two groups, depending on whether they consume whole pieces of plant tissue, as do the Colorado beetle, wheat midge and pea-weevil, or feed by sucking sap. The mouthparts of the sucking insects form a tube, through which sap can be withdrawn once the protective outer layer of the plant has been pierced. Such insects are often responsible for discolouring and deforming leaves, especially when they feed on young growth, and are also transmitters of virus diseases.

Many insects which feed in this way are bugs (Hemiptera): some are responsible for enormous losses in yield of agricultural and horticultural crops, especially the aphids (Aphidoidea), whiteflies (Aleyrodidae) and mealy bugs (Coccoidea). Among the aphids, one of the most serious pests is the peach aphid *Myzus persicae*, for it may feed on a wide variety of crops and transmits at least 24 different viruses. Another, which brought about a crisis in French vineyards at the end of the 19th century, is *Phylloxera vastatrix*. This species was accidentally introduced from America, where it forms galls on vine leaves, but it attacked the roots of the European vine *Vitis vinifera*. The problem was solved eventually by grafting European vines on to American stocks, a practice which has extended throughout the wine-growing areas of Europe. Aphids can reproduce rapidly in summer, often without fertilization, and the young are born alive. The tarnished plant-bug *Lygus rugulipennis* transmits a virus disease of oil-seed rape crops.

Many of these sucking insects move away from the arable fields in late summer and survive the cold part of the year in the shelter of neighbouring hedges or rough grassland. The adults of most species of aphid die in the autumn, after first laying fertilized eggs, but those of the peach aphid are able to live through the winter. Their chief winter host is the peach *Prunus persica*, but many other plants, woody or herbaceous, seem to serve almost as well.

Predator-prey relationships

Many ecologists consider that plant communities with a large number of species are more stable than those with only a few. Species-rich communities also support a greater variety of animals, including predators.

The extent to which predators can keep plant-eating insects in check depends on a number of factors, including the range of habitats tolerated by the two groups of animals. Three families of insects which prey on aphids are hover-flies (Syrphidae), lacewings (Chrysopidae) and ladybirds (Coccinellidae). Both larval and adult stages of ladybirds feed on aphids, but only the larvae of the other two. Some hover-flies are wide-ranging and likely to lay their eggs near aphid colonies in crops, while others are essentially woodland species and remain close to trees and shrubs. Green lacewings, such as *Chrysopa carnea*, and ladybirds, such as the two-spot *Adalia bipunctata*, occur in many habitats, including crops.

Often, in aphid epidemics, their predators increase to such an extent that the populations of aphids suffer catastrophic declines in summer and the numbers that survive to lay overwintering eggs are much less than in non-epidemic years. Usually, however, much damage is done to the crop before the collapse occurs, and farmers commonly resort to poisonous sprays which, unfortunately, also kill the predators. The use of pesticides can thus sometimes be counter-productive.

Of every 10,000 eggs of the large white butterfly *Pieris brassicae*, it is estimated that only 32 survive to the adult stage: all but about 400 are destroyed by insect parasites and predators and most of the remainder by vertebrates, particularly birds. If the crop is sprayed early in the year, many more caterpillars survive to maturity as a result of the destruction of these insect predators and parasites and the subsequent damage can be much more serious.

Similarly, the spraying of orchard crops against pests has had such an effect on the predators of red spider mites *Tetranychus/Panonychus*, among which the black-kneed capsid *Blepharidopterus angulatus* is perhaps the most important, that the mites themselves present a new problem: red spider mites seem to be relatively resistant to contact-pesticides and also breed rapidly, passing through several generations in the course of a single summer.

Herbivorous mammals

Wood mice *Apodemus sylvaticus*, brown hares *Lepus capensis* and rabbits *Oryctolagus cuniculus* are all considered to be serious pests of arable because of the damage they do to crops. The greatest amount by far is done by rabbits; before myxomatosis it was estimated that an average of 1½ cwt per acre (173 kg/hectare) of grain was lost each year over England and Wales as a result of rabbit attack. Whole fields of winter wheat were grazed so severely in winter that they were total failures. In contrast, damage done by wood mice is mostly restricted to the neighbourhood of hedges.

Unlike rabbits, hares are solitary creatures and furthermore live exclusively above ground. During the day they rest between grass tussocks or in depressions in the ground known as 'forms'. The 'mad' behaviour exhibited in February and March is the courtship display and is performed by the male or 'jack'. While rabbits are born naked, helpless and with their eyes closed, leverets have a short furry coat, their eyes are open and they can move about immediately after birth.

Birds of arable and farmsteads

Most of the birds that nest on the ground in grasslands will also breed in cereal crops. Nowadays, however, they are less likely to do well because of the relative frequency with which agricultural machinery is taken on to arable land to fertilize or spray. The skylark *Alauda arvensis* appears to be the most generally successful breeding bird on modern farms, quickly nesting again if it loses a brood to cultivation. The corn bunting *Miliaria calandra*, which sometimes nests in crops, generally escapes interference because it breeds late, from late May to the end of July. Populations of lapwings *Vanellus vanellus*, stone-curlews *Burhinus oedicnemus* and, on the Continent, little bustards *Tetrax tetrax* appear to be adapting to modern farming practices.

A bird that seems to favour ploughed fields, and is probably relatively successful because it usually nests in the rough grass around the edges, is the grey partridge *Perdix perdix*. Since the early 1960s, however, a marked decline in its numbers both in Britain and on the Continent, particularly in Denmark, appears to have been due to a combination of factors, some of which relate to the feeding of the chicks. Important in the diet of young partridges are the larvae of sawflies (Hymenoptera) that feed on cereal crops. These insects are commonest where the crop is undersown with clover ley, but a recent decline in this practice has increased the dependence of the chicks on other foods, especially aphids. The movements of aphids into cereal crops are affected by the climate and a long spell of cool weather in spring and early summer can delay these for several weeks. In such conditions, and

A leveret camouflaged in a ploughed field. Unlike rabbits, hares are born with a short, furry coat, open eyes, and are able to move around immediately.

there has been a tendency for the spring to be cool in recent years, many chicks die of starvation. Young red-legged partridges *Alectoris rufa* hatch later, and it is significant that they appear to have fared considerably better.

Several birds move into ripening cereal crops to feed on the grain. Notable among these are woodpigeons *Columba palumbus* and house sparrows *Passer domesticus*. Flocks of woodpigeons and rooks *Corvus frugilegus* eat newly-sown cereals, too. Rooks, however, also take insects regarded as agricultural pests, especially weevils (Curculionidae) and wireworms, the larvae of click-beetles (Elateridae), and it is difficult to judge whether or not, on balance, their feeding habits benefit the farmer.

The rook is as closely linked with agriculture as any European bird: not only does it feed on arable land, but it also often nests in small copses planted to provide shelter for farmsteads. Many woodland species also breed in these copses while other birds, such as the jackdaw *Corvus monedula*, swallow *Hirundo rustica*, house martin *Delichon urbica*, swift *Apus apus* and barn owl *Tyto alba*, nest on or inside the buildings. All these birds also nest in tree holes, in caves or on cliffs, sites which in many ways resemble those which have been provided in abundance by man.

Thus, just as the farmstead is the centre of activity for man, so is it for the bird-life of farming country. It provides not only a wide variety of micro-habitats to be exploited by the birds but an abundant supply of food as well. Pigeons, sparrows and finches feed on the grain, barn owls prey upon the brown rats *Rattus norvegicus* and house mice *Mus musculus* which are also attracted by the grain and other stored food, and swallows, house martins, swifts and other insectivorous birds feed on the flying insects associated with the domesticated animals, their dung or the farm pond. In recent years the increasing use of artificial cattle foods has attracted starlings *Sturnus vulgaris* in such large numbers in winter that they eat a significant proportion of the food and are thus regarded as serious pests. Every evening these birds congregate at communal roosts often consisting of tens of thousands of individuals. Plantations of spruce *Picea* are sometimes chosen and the droppings may accumulate on and beneath the trees to such an extent that they are killed. Before the coming of the combine harvester large numbers of seed-eating birds visited farmyards, while small mammals, including harvest mice *Micromys minutus*, congregated in corn stacks awaiting threshing.

One of the most remarkable happenings of the last few decades has been the colonization of central and western Europe by the collared dove *Streptopelia decaocto*. In 1930, the northern and western limits of its range were in the Balkans, but then began an explosive extension to the northwest across Europe. The Netherlands and Scandinavia were colonized and, in 1955, Britain. Collared doves are now common over most of lowland Britain and Ireland, and in the early 1970's they even bred in Iceland. They tend to avoid open country and are often found near farmsteads, where they take considerable quantities of grain, as well as in urban areas.

The Lowland Vegetation of Northern Europe
Arctic circle
20°
10°
0°
65°
60°
55°
50°
ICELAND
NE
WE
SCOTLAND
IRELAND
BRITAIN
WALES
ENGLAND
Forest of Dean
New Forest
NETHERLANDS
BELGIUM
Ardennes
LUXEMBOURG
FRANCE
R. Meuse
Enclosed
0 50 100 150 200 250 300 Miles
0 100 200 300 400 Kilometres

20°
30°
40°
Arctic circle
65°
60°
55°
50°
FINLAND
SWEDEN
WAY
NMARK
R. Oder
Kampinos Forest
Białowieża Forest
POLAND
EAST GERMANY
Świętokrzyzie (Holy Cross Forest)
R. Oder
CE
Thüringer Wald
CZECHOSLOVAKIA
ST GERMANY
Predominantly lowland grassland
Predominantly arable
Mixed farmland (arable and grassland)
Major forests (entirely or partially broadleaved)
Boundary separating enclosed from non-enclosed land
200m contour
NE North Europe
WE West Europe
CE Central Europe

GLOSSARY

abdomen in insects, spiders, posterior section of body, lacking appendages
acute ending in a point
alternate leaves placed singly at different positions along the stem; *cf* opposite
anther part of stamen producing pollen
apex tip or summit; wing tip in insects
-ate describing 2-dimensional shape, *eg* ovate
awn in grasses, long, stiff bristle projecting beyond grain
axil angle between leaf and stem
basal in plants, at base of stem; in animals, position nearest body
bract modified leaf, often at base of flower-stalk
bulbil small bulb arising in leaf axil or replacing flower; can grow into new plant
calcareous of soil, overlying chalk or limestone, therefore alkaline
calyx, calyces (plural) all the sepals; term often used when the sepals are joined to form a tube
cerci projections from rear of abdomen
ciliate edged with hairs
dextral in snails, shell with right-handed spiral; *cf* sinistral
femur, femora (plural) in insects, basal part of leg
gills in fungi, thin wing-like structures below cap, producing spores
hermaphrodite in animals, ♂ and ♀ organs in same individual
inflorescence flower branch, including bracts, flower-stalks and flowers
lanceolate spear-shaped
linear long and narrow
mantle in slugs, snails, skirt of tissue enclosing viscera, lungs; visible as dorsal shield of skin in slugs
mesonotum dorsal centre of thorax
metamorphosis change of form found in insects, amphibians. In insects, complete metamorphosis: young radically different from adult, involves a pupal stage, *eg* caterpillar to butterfly; incomplete metamorphosis: young similar to adult, no pupal stage, *eg* aphid
microspecies groups of individuals differing only slightly from other groups of individuals but not interbreeding with them
nerve strand of strengthening or conducting tissue running through leaf
ob- inverted, with broadest part of structure near apex, *eg* obovate
-oid describing 3-dimensional shape, *eg* ovoid
opposite leaves paired on opposite sides of stem; *cf* alternate
ovate, ovoid egg-shaped (see -ate, -oid)
ovipositor egg-laying apparatus of ♀ insects; in ☿ bees and wasps modified into sting
panicle branched inflorescence
pinnate regular arrangement of leaflets in 2 rows on either side of stalk (simply p.), each leaflet divided again (twice p.)
perianth segment one floral leaf, used when petals and sepals are indistinguishable
peristome in snails, outer lip of shell aperture
proboscis in insects, tubular mouthpart
pronotum dorsal front of thorax
rostrum in insects, snout or beak
saprophyte plant deriving its food totally or partially from dead organic matter
scutellum in bugs, triangular section in centre of thorax
sepals outer ring of floral leaves, usually green and less conspicuous than petals
simple leaf not divided into segments
sinistral in snails, shell with left-handed spiral; *cf* dextral
spathe enlarged bract surrounding flowerhead
species group of similar individuals which can interbreed; cannot usually interbreed with other species to produce fertile offspring
spikelet in grasses, one or more florets enclosed by a pair of stiff bracts
spire in snails, shell above most recently formed whorl (*qv*)
spore in plants, minute reproductive body, produced asexually and of simpler structure than seed
stamen male organ of flower
stigma part of female organ of flower which receives pollen
stipule scale at base of leaf-stalk

style female organ of flower
subspecies group of individuals within a species having distinctive features
superciliary in birds, stripe above eye
suture in snails, groove between successive shell whorls
thallus plant body undifferentiated into leaf, stem, etc; often flattened
tibia in insects, middle section of leg
thorax in insects, spiders, region between head and abdomen, bearing legs, wings
umbel cluster of flowers whose stalks (rays) radiate from top of stem
umbilicus in snails, hollow base of coiling axis of spiral shells
variety group of individuals within a species having one or more distinctive characteristics; less differentiated from other members of species than a subspecies
viviparous giving birth to live young
whorls leaves or flowers arising in circles around stem; in snails, sections of shell

ABBREVIATIONS

The ranges in the order of their listing in the field guide

W	widespread
T	throughout
Br	Britain (England, Scotland, Wales)
Ir	Ireland
Ic	Iceland
Fr	France, north of the Loire
Lu	Luxembourg
Be	Belgium
Ne	Netherlands
De	Denmark
Ge	Germany
Cz	Czechoslovakia
Po	Poland
Fi	Finland
Sw	Sweden
No	Norway
FS	Fenno-Scandia (Norway, Sweden, Finland)
SC	Scandinavia (Norway, Sweden)
NE	Fenno-Scandia, Denmark, north Germany, north Poland
CE	Czechoslovakia, south Germany, south Poland
WE	Britain, Ireland, France, Luxembourg, Belgium, Netherlands

n, s, e, w, c north, south, east, west, central

When the species is not native but introduced and naturalized, the countries are put in brackets, *eg* Fr, Ge, (Br, Ir).

c	about
av	average
esp	especially
fld	flowered
fl(s)	flower(s)
fl-head	flowerhead
fr(s)	fruit(s)
imm	immature
inflor	inflorescence
juv	juvenile
lf (lvs)	leaf (leaves)
lflet	leaflet
lfy	leafy
microsp(p)	microspecies
sp	species (singular)
spp	species (plural)
ssp	subspecies
var	variety

MEASUREMENTS

Scale in the plates: the relative sizes of the plants and animals are preserved whenever possible, but the measurements in the entries themselves should be noted.

BL	body length (excludes tail in mammals, includes antennae in insects)
EL	extended length (slugs); ear length (mammals)
FA	forearm length
H	height
HF	hindfoot length
L	total length (includes beak, tail)
SB	shell breadth
SH	shell height (snails); shoulder height (mammals)
TL	tail length
W	number of whorls
WS	wingspan

SYMBOLS

♀	female
♂	male
☿	worker
<	up to
>	more than
?	doubtful
U	underside

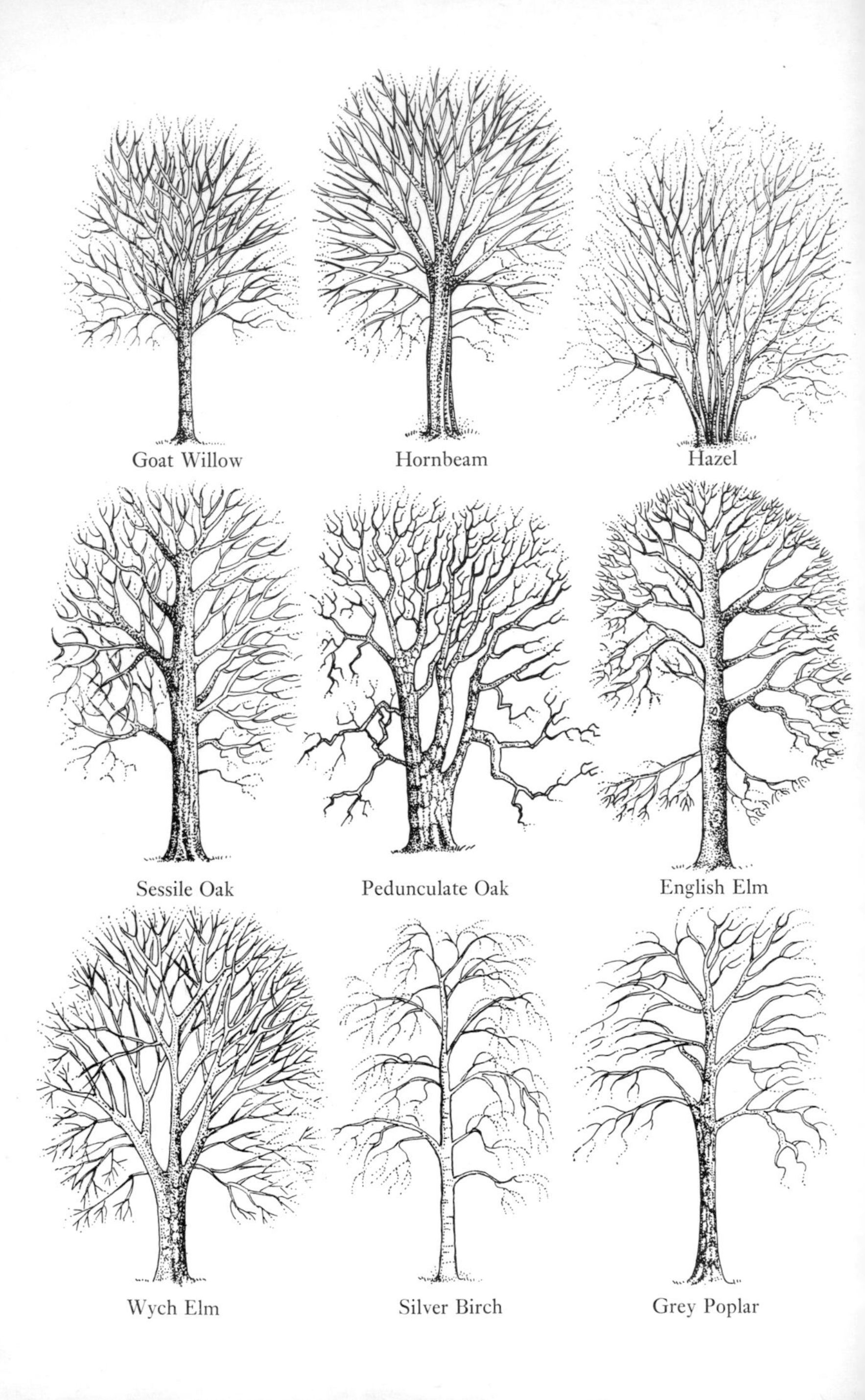
Goat Willow
Hornbeam
Hazel
Sessile Oak
Pedunculate Oak
English Elm
Wych Elm
Silver Birch
Grey Poplar

Hawthorn
Crab Apple
Beech
Wild Cherry
Field Maple
Sycamore
Ash
Small-leaved Lime
Common Whitebeam

lichen *Cladonia pocillum* CLADONIACEAE
H 1–2 cm. Continuous layer of small, grey green 'leaves' like crazy paving. Spores produced in funnel-shaped, stalked cups covered in warts. Calcareous grassland, walls, rocks. T. [1]

lichen *Graphis scripta* GRAPHIDACEAE
White to grey with black slit-like, spore-producing areas. Smooth-barked trees where pollution low. T, ex Ic. [2]

moss *Dicranum scoparium* DICRANACEAE
H 2–3 cm. Lvs <5 mm, long and narrow, all turned to one side, with long, jagged-toothed points into which nerves run. Spore capsules cylindrical, curved, with long-beaked lids, on stalks <3 cm. Woods, heaths. T. [3]

Fawley Bun *Leucobryum glaucum*
DICRANACEAE H 5–20 cm. Lvs <1 cm, lanceolate, with broad, thick nerve to ½ way. Spore capsules rare, small, cylindrical, drooping, furrowed when dry. Easily recognized by whitish-green of dead cushions. Woods, heaths. T, ex Ic. [4]

moss *Fissidens taxifolius* FISSIDENTACEAE
H 10–15 mm. Creeping oblong-lanceolate lvs with clasping pocket at base, nerve extending from tip as stout point. T, ex Ic. [5]

Cord-moss *Funaria hygrometrica*
FUNARIACEAE H 2–7 cm. Lvs <5 mm, oval-oblong; upper clustered into tuft, nerved almost to tip. Spore capsules pear-shaped, on long stalks, orange or yellowish-brown, with beautifully marked lids. Woods, heaths, esp after fires. T. [6]

moss *Mnium undulatum* BRYACEAE
H 2–10 cm. Palm-like, forming large, loose patches. Lvs <15 mm, increasing up stem, tongue-shaped, blunt-toothed, with nerve exceeding apex, wavy margins. Spore capsules oval, drooping, on long stalks <3 cm. Moist woods. T. [7]

moss *Thamnium alopecurum* LEUCODONTACEAE H 7–20 cm. Stems creeping or erect, divided into curved branches turned to one side. Lvs on upright stems broadly triangular, those on branches

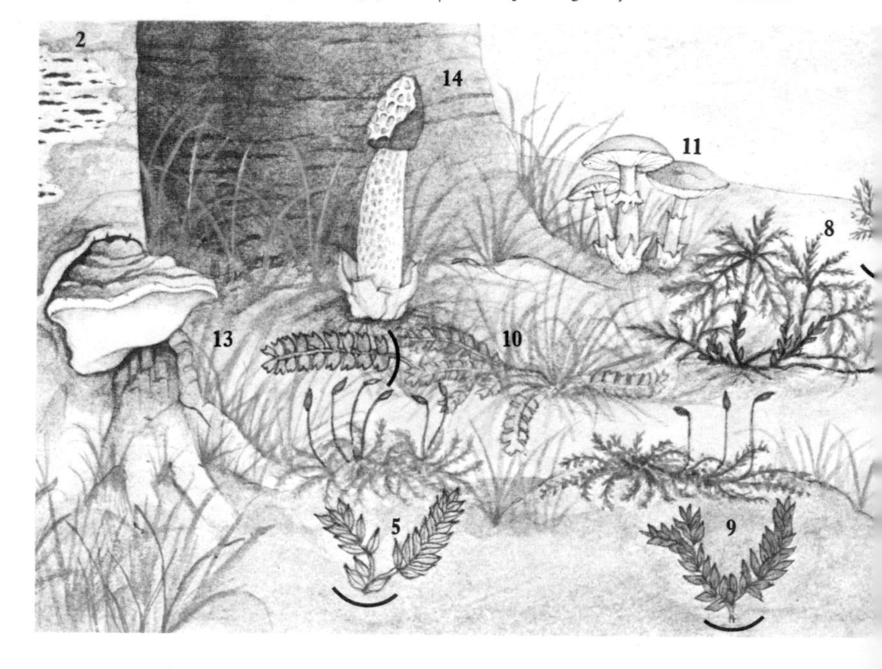

oblong. Spore capsules oblong, drooping on short stalks <12 mm. Woods, waterfalls. T. [8]

moss *Plagiothecium denticulatum* HYPNACEAE Creeping, with flattened, shiny shoots. Lvs <3 mm, oval, with sharp-pointed tips, double-nerved almost to middle. Spore capsules cylindrical, distinctly necked, almost upright on orange-red stalks <4 cm. Woods. T, ex Ic. [9]

liverwort *Lophocolea bidentata* JUNGERMANNIACEAE Creeping, leafy, in loose, pale green patches. Lvs nerveless, with 2 teeth at tip, concealing 4-lobed under-lvs. Spore capsules rare, brown, oblong-oval, upright on long stalks. Damp grass. T, ex Ic. [10]

Death Cap *Amanita phalloides* AGARICINEAE H 7–12 cm. Cap greenish or yellowish-olive, somewhat slimy, <9 cm across. Stalk white with soft white ring above, remains of second ring (volva) below. Gills white. May–Nov. Woods, pastures. Poisonous. T, ex Ic. [11].

Honey Fungus *Armillarea mellea* AGARICINEAE H 7–15 cm. Cap yellowish to brown with dark, honey-coloured, hairy scales, <15 cm across. Gills cream to yellowish-brown, Aug–Nov. In dense clusters at bases of old trees. T. [12]

Tinder Fungus *Ganoderma applanatum* POLYPORACEAE Semicircular, <25 cm across, on living trees, esp beech; upper surface red-brown; lower surface of tiny pores, at first white, becoming red-brown with age. All year. T, ex Ic. [13]

Stink-horn *Phallus impudicus* PHALLOIDACEAE H <20 cm. Egg-shaped fungus, <5 cm across; when ripe pushes up white stalk bearing cap covered in olive-green, spore-producing jelly. Vile smell attractive to flies which eat and distribute spores. May–Nov. Edible before expansion. T, ex Ic. [14]

The structures drawn within each semi-circle are magnified

Giant Puff-ball *Calvatia gigantea* LYCOPERDACEAE H <30 cm. White or yellow-tinged, football-sized fungus, olive-brown when old and dry. Breaks open to release *c*7 million spores. Jun–Oct. Woods, pastures. Edible when young, fried in slices like bread. T, ex Ic. [1]

Field Horsetail *Equisetum arvense* EQUISETACEAE H 10–80 cm. Erect, perennial herb with creeping, underground stems, producing 2 kinds of hollow aerial stems, the first to appear being pale brown, fertile, without branches, followed by others, green, sterile. Spores ripe Apr, in terminal ovoid cones <4 cm. Hedge-banks, arable fields. T. [2]

Wood Horsetail *Equisetum sylvaticum* EQUISETACEAE H 10–80 cm. Erect, perennial herb with creeping, underground stems; fertile stems shorter, <40 cm, with terminal cones <25 mm. Spores ripe Apr–May. Damp, acid woods. T. [3]

Great Horsetail *Equisetum telmateia* EQUISETACEAE H 50–200 cm. Robust, erect, perennial herb with creeping, underground stems, producing 2 kinds of hollow aerial stems, the first to appear being pale brown, fertile, without branches, followed by others, dirty white, sterile, with spreading branches. Spores ripe Apr, in terminal, ovoid cones <8 cm. Damp woods, banks. T, ex Ic, Fi, No. [4]

Bracken *Pteridium aquilinum* HYPOLEPIDACEAE H <2·5 m. Perennial herb with far-creeping, underground stems. Sporangia (fruiting bodies which produce spores) covered by rolled lf-margin. Spores ripe Jul–Aug. Woods; also heaths, moors. T, ex Ic.

Hart's-tongue *Phyllitis scolopendrium* ASPLENIACEAE H 10–60 cm. Lvs undivided, strap-shaped. Sori (clusters of sporangia) linear, in pairs across lf. Spores ripe Jul–Aug. Hedges, rocky woods. T, ex Ic, Fi.

Lady-fern *Athyrium filix-femina* ATHYRIACEAE H 20–100 cm. Perennial herb with short, stout, underground stems. Sori oblong, in 2 rows down lf-segments, covered by flap. Spores ripe Jul–Aug. Woods, hedge-banks. T. [5]

Brittle Bladder-fern *Cystopteris fragilis* ATHYRIACEAE H 5–35 cm. Tufted, perennial herb. Sori in 2 rows down lf-segments, covered by flap. Spores ripe Jul–Aug. Rocky woods, walls. T. [6]

Male-fern *Dryopteris filix-mas* ASPIDIACEAE H 30–130 cm. Lf-stalk ½ as long as lf, scaly. 3–6 circular sori on undersides of each lf-segment. Spores ripe Jul–Aug. Rocks, woods, hedges, walls. T.

Hard Shield-fern *Polystichum aculeatum* ASPIDIACEAE H 30–90 cm. Tufted, perennial herb, somewhat leathery. Sori circular, in row on each side of midrib. Spores ripe Jul–Aug. Woods, hedge-banks. T, ex Ic, De, Fi. [7]

Broad Buckler-fern *Dryopteris dilatata* ASPIDIACEAE H 10–150 cm. Tufted, perennial herb with erect, underground stem. Lf-stalks with dark, brown-centred scales. Sori circular, in row on each side of midrib. Spores ripe Jul–Sep. Woods, hedgerows, damp rocks. T, ex Ic. [8]

Limestone Fern *Gymnocarpium robertianum* ASPIDIACEAE H 15–60 cm. Perennial herb with long, creeping, underground stems. Lvs solitary, glandular hairy, dull green, when young rolled up to form single ball. Sori close to edges of segments, not covered by flap. Spores ripe Jul–Aug. Limestone rocks, screes. T, ex Ic, De. [9]

Polypody *Polypodium vulgare* POLYPODIACEAE H 10–30 cm. Lf-stalks ⅓–¾ as long as lf. Two rows of oval to circular sori on undersides of each lf-segment. Spores ripe in spring. Rocks, hedge-banks, walls; often grows on trees. T.

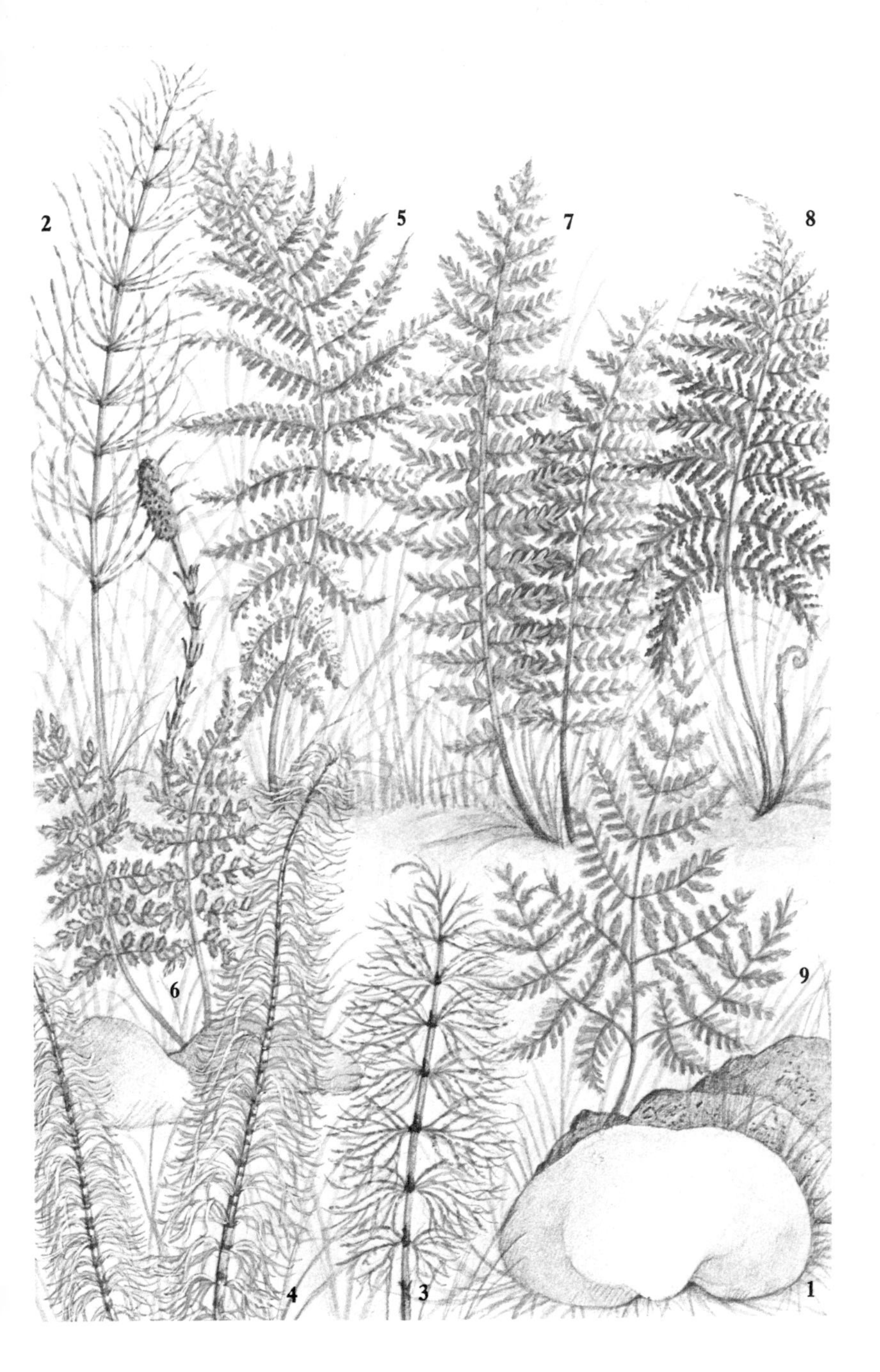
2
5
7
8
6
9
4
3
1

Silver Birch *Betula pendula* BETULACEAE
H <30 m. Medium-sized, silver-barked tree, with pendulous, hairless, warty twigs. Lvs triangular, prominently double-toothed. Fls catkins, ♂ 3–6 cm, Apr–May. Open woodland, quarries. Supports many insect spp, esp moth caterpillars. T, ex Ic. [1]

Fluttering Elm *Ulmus laevis* ULMACEAE
H <35 m. Deciduous tree with grey bark falling in thin flakes. Lvs <8 cm, ovate, short-stalked, with curly hairs beneath. Fls Mar–Apr. Frs flattened, with hairy margins, notched top. Fls and frs on long stalks, 'flutter' in the wind. Damp woods. Fr, Be, Ge, Cz, Po, Fi, Sw. [2]

Wych Elm *Ulmus glabra* ULMACEAE
H <40 m. Broad, domed, deciduous tree with ascending branches. Lvs oblong, rough to touch above, occasionally 3-pointed. Fls before lvs, Feb-Mar. Frs dish-like, with seed in centre, May–Jun. Woods, hedgerows. T, ex Ic. [3]

English Elm *Ulmus procera* ULMACEAE
H <30 m. Erect, deciduous tree with few, large, heavy branches. Lvs smaller, more rounded than wych elm, also hairier and less rough above, longer petioles. Fls Feb–Mar. Frs May–Jun, but rarely produce viable seed. Hedgerows. Br, Fr. [4]

Goat Willow *Salix caprea* SALICACEAE
H <10 m. Deciduous shrub or small tree with stout twigs, hairy when young. Lvs 5–10 cm, broadly ovate, with blunt tips, softly hairy beneath. Fls catkins, appearing before lvs, ♀ larger than ♂, Mar–Apr. Woods, hedges. Bark once used to make aspirin. T, ex Ic. [5]

Grey Poplar *Populus canescens* SALICACEAE
H <35 m. Deciduous tree with spreading crown. Lvs <10 cm, nearly round, broadest below middle, with 4–6 teeth on each side, grey-hairy beneath (hence name). Fls Feb–Mar. Damp woods, often in groups spread by suckers. Frequent host of mistletoe *Viscum album*. Br, Fr, Ne, Ge, Cz, Po, (Ir, Be, De, Sw). [6]

Mistletoe *Viscum album* LORANTHACEAE
H <1 m. Woody evergreen perennial, with smooth, green, much-branched stems. Semi-parasitic on trees. Lvs <8 cm, yellowish-green, leathery, opposite, narrowly ovate, blunt, on short stalks. Fls green, with 4 petals, small, in compact clusters of 3–5; ♂ and ♀ on separate plants, ♂ lacks calyx; Feb–May. Frs white, spherical berries, <1 cm. On broad-leaved trees, particularly poplar *Populus*, apple *Malus*, lime *Tilia*, willow *Salix*, and hawthorn *Crataegus*; found on fir *Abies* only in Po, Cz. T, ex Ic, Fi, (but Ir). [7]

Hornbeam *Carpinus betulus* CORYLACEAE
H 30 m. Deciduous tree with fluted trunk, smooth grey bark, ascending branches. Lvs <10 cm, oval, strongly veined, sharply double-toothed. Fls catkins, ♂ <5 cm, Apr–May. Frs <2 cm, half-enclosed by 3-lobed wing. Woods. T, ex Ir, Ic, Fi, No. [8]

Hazel *Corylus avellana* CORYLACEAE
H <6 m. Deciduous shrub, branched from ground, with smooth, brown bark, hairy twigs. Lvs <12 cm, round, hairy on both sides, with double-toothed margins. Fls catkins, appearing before lvs, ♂ >8 cm, conspicuous, ♀ small, bud-like, with protruding red styles, Feb–Mar. Frs edible nuts. Woods, hedges. T, ex Ic. [9]

Sessile Oak *Quercus petraea* FAGACEAE
H <40 m. Deciduous tree with upward-spreading branches, deeply fissured bark. Lvs <12 cm, with 5–8 pairs of rounded lobes, wedge-shaped bases. Fls catkins, appearing with lvs, ♂ <4 cm, Apr–May. Frs stalkless acorns. Woods, hedges, on acid soils. T, ex Ic, Fi. [10]

Pedunculate Oak *Quercus robur* FAGACEAE
H <45 m. Large, deciduous tree with massive trunk, spreading branches. Lvs lobed, heart-shaped at base, on very short petioles. Fls catkins, ♂ 2–4 cm, Apr–May. Frs acorns, on long stalks 2–8 cm, Sep–Oct. Woods, hedges. Supports more spp of insects than any other tree. T, ex Ic. [11]

5
8
9
10
11
3
2
6
7
4
1

Beech *Fagus sylvatica* FAGACEAE H <40 m. Large deciduous tree with smooth, grey bark. Lvs ovate, pointed, with wavy margins. ♂ catkins on slender stalks, 5–6 cm, Apr–May. Frs triangular, in prickly husks, Sep–Oct. T, ex Ic, Fi, (but Ir). [1]

Hop *Humulus lupulus* CANNABACEAE H <6 m. Perennial herb, climbing clockwise. Lvs <15 cm, opposite, on long stalks, deeply 3–5 lobed, coarsely toothed. ♂ and ♀ fls on separate plants, Jul–Aug: ♂ stalked, perianth and 5 stamens; ♀ covered by bracts, clustered, developing into cone-like frs <5 cm, used in brewing. Damp woods, hedges. T, ex Ic. [♀ 2]

Small Nettle *Urtica urens* URTICACEAE H 10–60 cm. Annual herb with stinging hairs. Lvs <4 cm, opposite, elliptical, strongly toothed, on short stalks. Fls green, ♂ and ♀ separate in spike-like clusters, Jun–Sep. Cultivated ground, and waste places. T. [3] Common nettle *U. dioica* is perennial, <2 m, with ♂ and ♀ fls on separate plants. Woods, hedges, nitrogen-rich grassland, waste places. T.

Pale Persicaria *Polygonum lapathifolium* POLYGONACEAE H <80 cm. Very variable, erect, hairy, annual herb with greenish stems swollen beneath each lf-base. Lvs <20 cm, lanceolate, often black-blotched. Fls greenish-white, in stout, obtuse spikes, covered in yellow glands, Jun–Oct. Arable fields, gardens. T, ex Ic. [4] Redshank *P. persicaria*, in similar habitats, but fls pink and without glands. T.

Japanese Knotweed *Reynoutria japonica* POLYGONACEAE H <2 m. Erect, perennial herb, sending up numerous, hairless, hollow stems to form dense thicket. Lvs <12 cm, broadly oval, long, pointed, with square-cut bases. Fls white, in clusters of 2–4 on loose branches, Aug–Sep. Waste places, rivers, stream banks; escape from cultivation first established in Britain in S Wales, 1887, now ubiquitous. (T, ex Ic, De, Sw, from China, Japan) [5]

Broad-leaved Dock *Rumex obtusifolius* POLYGONACEAE H 60–100 cm. Erect, hairless, branched, perennial herb. Lvs <25 cm, broadly oblong, with heart-shaped base, obtuse apex and wavy margins. Fls of 3 small, 3 large, toothed segments, only one of latter developing tubercle (reddish swelling containing fr); in whorls and loose spikes, Jun–Oct. Waste places, arable land. T. [6] Curled dock *R. crispus*, similar in habitat and distribution, though not Ic, has longer, lanceolate lvs <30 cm; larger segments of fls lack teeth and all 3 have tubercles.

Many-seeded Goosefoot *Chenopodium polyspermum* CHENOPODIACEAE H 15–100 cm. Erect, hairless, annual herb with 4-angled stem, often red. Lvs <8 cm, oval to elliptical, toothless (or, rarely, with single tooth on each side above base). Fls tiny, green, in long, loose, branched spike, Jul–Oct. Frs with numerous black seeds. Waste places, arable fields. T, ex Ic. [7]

Fat-hen *Chenopodium album* CHENOPODIACEAE H 30–90 cm. Erect, mealy, annual herb, stem often red. Lvs variable in shape, often long, rhomboid and toothed, but always mealy-white when young, green later. Fls small, in greenish-white clusters. Arable fields, waste places. Seeds and leaves formerly eaten. T, ex Ic. [8]

Common Orache *Atriplex patula* CHENOPODIACEAE H <150 cm. Much-branched, slightly mealy, annual herb with erect, spreading or prostrate, ridged stems. Lvs <10 cm, lower rhomboid with wedge-shaped bases, upper lanceolate. Fls tiny, green, ♂ and ♀ separate, Jul–Oct. Frs enclosed in 2 persistent, lf-like flaps, fused at base. Cultivated ground, waste places. T. [9]

Spear-leaved Orache *Atriplex hastata* CHENOPODIACEAE H 20–100 cm. Spreading or erect, annual herb, often with reddish stems branching from base. Lvs <10 cm, triangular, the margins of some making right-angle to stalk. Fls green, ♂ and ♀ separate, Jul–Sep. Frs enclosed in flaps fused at base. Roadsides, waste places. T, ex Ic.

2
1
5
6
8
7
3
4
9

Three-nerved Sandwort *Moehringia trinervia* CARYOPHYLLACEAE H <10 cm. Weakly trailing, hairy, perennial herb. Lvs <25 mm, ovate, strongly 3-veined, lower stalked. Fls white, <6 mm, with 5 petals only ½ as long as 3-veined sepals, May–Jul. Woods, hedge-banks, shady walls. T, ex Ic. [1]

Common Chickweed *Stellaria media* CARYOPHYLLACEAE H 5–40 cm. Low-growing, annual herb. Lvs <25 mm, opposite, elliptical, short-stalked. Fls <8 mm, white, with 5 strongly divided petals, all year. Arable land, waste places. T.

Greater Stitchwort *Stellaria holostea* CARYOPHYLLACEAE H 10–20 cm. Ascending perennial herb with rough, brittle, quadrangular stems. Lvs >8 cm, opposite, narrowly lanceolate, tapering to long, fine points. Fls <30 mm, white, with 5 petals divided to ½ way; in loose, branched clusters, May–Sep. Woods, hedgerows. T, ex Ic. [2]

Sticky Mouse-ear *Cerastium glomeratum* CARYOPHYLLACEAE H 5–45 cm. Erect, sticky, yellowish-green, annual or overwintering herb. Lvs >25 mm, opposite, ovate (upper broadly so), all stalkless, hairy. Fls <5 mm, white, with 5 petals as long as calyx divided ¼ way, in dense clusters, Apr–Sep. Cultivated ground, waste places. T, ex Fi. [3]

Common Mouse-ear *Cerastium holosteoides* CARYOPHYLLACEAE H 25–30 cm. Prostrate, somewhat hairy, perennial herb. Lvs 10–25 mm, opposite, blunt, without stalks. Fls <15 mm, white, with 5 notched petals, usually no longer than sepals, arranged on regularly branched stems, Apr–Sep. Grassland, roadsides. T.

Annual Knawel *Scleranthus annuus* CARYOPHYLLACEAE H <25 cm. Spreading, annual herb with branched, ascending, often hairy stems. Lvs <25 mm, opposite, very narrow, grooved. Fls <4 mm, green, with 5 sepals, no petals,

in axils and in terminal clusters, Jun–Aug. Dry, sandy places, cultivated ground. T, ex Ic. [4]

Corn Spurrey *Spergula arvensis* CARYOPHYLLACEAE H 5–70 cm. Erect, but weak-stemmed, glandular-hairy, annual herb. Lvs <3 cm, opposite, much-divided into linear segments, grooved beneath. Fls <7 mm, white, with 5 petals slightly exceeding sepals, Jun–Sep. Cultivated ground, esp sandy soils. T. [5]

White Campion *Silene alba* CARYOPHYLLACEAE H 30–100 cm. Erect, hairy, much-branched, perennial herb; often short-lived. Lvs <10 cm, opposite, ovate to lanceolate, stalkless. Fls <30 mm, white, with 5 petals divided to $\frac{1}{2}$ way, on separate ♂ and ♀ plants, May-Sep. Hedgerows, cultivated land. T, ex Ic. [6]

Red Campion *Silene dioica* CARYOPHYLLACEAE H 30–90 cm. Erect, hairy, much-branched, perennial or biennial herb. Lvs <10 cm, opposite, ovate, lower narrowed into long, winged stalks. Fls <25 mm, bright rose, with 5 petals divided to $\frac{1}{2}$ way, on separate ♂ and ♀ plants, May–Jun. Woods, shady hedgerows. T. [7]

Bladder Campion *Silene vulgaris* CARYOPHYLLACEAE H 25–90 cm. Bluish-green, perennial herb. Lvs <45 mm, opposite, elliptical to ovate, lower shortly stalked. Fls <18 mm, white, with 5 petals divided to $\frac{1}{2}$ way, ♂ and ♀ fls on same or on separate plants, Jun–Aug. Calyces form bladders surrounding frs. Grassland, roadsides, waste places. T. [8]

Soapwort *Saponaria officinalis* CARYOPHYLLACEAE H 30–90 cm. Branched, erect, hairless, perennial herb with long, underground stems. Lvs <10 cm, opposite, broadly elliptical, acute, stalkless, strongly 3-veined. Fls <25 mm, pink, with 5 unnotched petals, in compact, terminal clusters, Jul–Sep. Fr, Lu, Be, Ge, Cz, Po, (Br, Ir, De, FS). [9]

Green Hellebore *Helleborus viridis* RANUNCULACEAE H 20–40 cm. Tall, hairless, perennial herb with short, black, ascending stock, not overwintering. Lvs <20 cm across, divided into 5–7 narrow, toothed segments. Fls <5 cm, yellow-green, with 5 sepals, no petals, in branched, lfy cluster, Mar–Apr. Woods, scrub; also grown in gardens. Very poisonous. Br, Fr, Lu, Be, Ge, (Cz, Po). [1]

Monk's-hood *Aconitum napellus* RANUNCULACEAE H 50–100 cm. Stout, erect, leafy, perennial herb with tuberous stock. Lvs <15 cm across, spirally arranged, divided into 3–5 primary segments. Fls <2 cm long, bluish-mauve, with 5 outer perianth segments, uppermost forming helmet-shaped hood, May–Jun. Wet woods, shady stream-banks; also gardens, often escaping. Extremely poisonous. Br, Fr, Lu, Be, Ge, Sw. [2]

Wood Anemone *Anemone nemorosa* RANUNCULACEAE H 6–30 cm. Creeping perennial herb, spreading by brown underground stems, forming large patches. Lvs <5 cm across, divided into 3–5 lobes, each further cut into toothed, acute segments. Fls <4 cm, white or pink, with 5–9 perianth segments, Mar–May. Woods, shady places. T, ex Ic. [3]

Yellow Anemone *Anemone ranunculoides* RANUNCULACEAE H 7–30 cm. Similar to *A. nemorosa*, but with very short-stalked stem lvs, deeply 3-lobed, in whorl. Fls <2 cm, with 5 perianth segments, Apr. Woods. T, ex Br, Ir, Ic. [4]

Pasqueflower *Pulsatilla vulgaris* RANUNCULACEAE H 10–30 cm. Tufted, perennial herb with stout stock. Lvs <20 cm across, 2–3 times divided into long, narrow segments, silky-haired when young. Fls <85 mm, purple, usually with 6 'petals', solitary, bell-shaped, Mar–May. Frs 1-seeded developing silky plume <5 cm. Meadows, dry grassland. Poisonous. Br, Lu, Be, Ne, De, Ge, Cz, Po, Sw. [5]

Traveller's-joy *Clematis vitalba* RANUNCULACEAE H <30 m. Deciduous, woody climber. Lvs <20 cm, opposite, divided into 3–5 ovate, usually toothed lflets. Fls <2 cm, greenish-white, with 4 hairy 'petals', in terminal or axillary clusters, Jul–Aug. Frs 1-seeded, developing long, silky plume <4 cm to form 'old-man's-beard' (another name), which remains on plant throughout winter and into following spring. Woods, scrub, hedgerows. Br, Fr, Lu, Be, Ne, Ge, Cz, (Ir, Po, Sw, No). [6]

Lesser Celandine *Ranunculus ficaria* RANUNCULACEAE H 5–25 cm. Perennial herb with numerous, cylindrical root-tubers. Lvs <4 cm, heart-shaped, fleshy, shining dark green, on long stalks. Fls <3 cm, yellow, with 8–12 petals, solitary, Mar–May. Woods, scrub, hedgerows. Lvs, tubers poisonous. T, ex Ic. [7]

Meadow Buttercup *Ranunculus acris* RANUNCULACEAE H 15–100 cm. Erect, hairy, much-branched, perennial herb. Lvs long-stalked, roundish in outline, divided into 2–7 toothed segments. Fls 18–25 mm, yellow, on smooth stalks, with 5 petals, upright sepals, May–Jul. Meadows, roadsides. T. [8]. Bulbous buttercup *R. bulbosus*, like *R. acris*, but corm-like tuber at base of stem just below ground; fls with grooved stalks, reflexed sepals. T, ex Ic. Creeping buttercup *R. repens*, like *R. acris*, but spreading by extensive runners; fls with grooved stalks, upright sepals. T.

Goldilocks Buttercup *Ranunculus auricomus* RANUNCULACEAE H 10–40 cm. Perennial herb. Lower lvs long-stalked, roundish, coarsely toothed; upper deeply 3-lobed. Fls <25 mm, yellow, with 5 or fewer petals, on hairy, not furrowed stalks, Apr–May. Woods. T. [9]

White Buttercup *Ranunculus aconitifolius* RANUNCULACEAE H 20–50 cm. Erect, perennial herb, much-branched above. Lvs deeply divided into 3–5 lanceolate, toothed segments. Fls <2 cm, white, with 5 petals, arranged in loose, spreading clusters, May–Aug. Woods, meadows. Fr, Ge, Cz. [10]

6
9
2
10
1
3
4
7
8
5

Common Poppy *Papaver rhoeas*
PAPAVERACEAE H 20–60 cm. Erect, annual herb, often with stiffly hairy stems. Lvs <15 cm, 1–2 times divided into narrow, acute segments; lower stalked. Fls <10 cm, scarlet, often with black spot at base of 4 petals, Jun–Aug. Frs globular capsules, <2 cm. Arable land. T, ex Ic. [**1**]

Long-headed Poppy *Papaver dubium*
PAPAVERACEAE H 20–60 cm. Erect, branched, annual herb with spreading hairs on lower ½ of stem. Lvs like *P. rhoeas*, but terminal segments smaller than laterals. Fls pale scarlet, sometimes with black spot at base, Jun–Jul. Frs capsules, <2 cm, 2–3 times as long as broad. Arable land. T, ex Ic. [fr only **2**]

Prickly Poppy *Papaver argemone*
PAPAVERACEAE H 15–45 cm. Erect, branched, annual herb with stiff, appressed hairs. Lvs <15 cm, deeply 1–2 times divided into linear segments. Fls <6 cm, scarlet, with dark base, Jun–Jul. Frs capsules, <25 mm, 2–3 times as long as broad, strongly ribbed, with erect bristles. Arable land. T, ex Ic. [fr only **3**]

Greater Celandine *Chelidonium majus*
PAPAVERACEAE H 30–90 cm. Erect, bluish-green, sparsely hairy, perennial herb, exuding orange latex when broken. Lvs <25 cm, divided into 5–7 oblong, toothed lflets. Fls <25 mm, yellow, with 4 petals, May–Aug. Frs capsules, <5 cm long, narrow. Banks, hedgerows. T, ex Ic. [**4**]

Tuberous Corydalis *Corydalis solida*
PAPAVERACEAE H 10–20 cm. Erect, perennial herb with solid, rounded, tuberous rootstock; 1–2 scale lvs below true lvs. Lvs <25 mm, divided into 3 deeply cut segments. Fls <25 mm, dull purple, with long spur, in dense, terminal clusters, Mar–May. Frs 2-valved capsules. Woods, hedgerows. Fr, Lu, Be, Ne, Ge, Cz, Po, Fi, Sw, (Br, De, No). [**5**]

Common Fumitory *Fumaria officinalis*
PAPAVERACEAE H 20–70 cm. Slender, erect, much-branched, annual herb. Lvs <10 cm, much-divided into flat, linear segments. Fls <9 mm, pink, with blackish-red petal tips, Apr–Sep. Frs spherical, with apical notch, <2·5 mm. Arable fields, waste places. T, (but Ic). [**6**]

Hedge Mustard *Sisymbrium officinale*
CRUCIFERAE H 30–90 cm. Erect, bristly, annual herb. Lvs <8 cm, deeply divided, with large, round, terminal lobe. Fls <3 mm, with 4 yellow petals, in spikes elongating in fr, Jun–Jul. Frs pods, <15 mm, hairy. Hedge-banks, walls, waste places, arable fields. T, ex Ic. [**7**]

Tall Rocket *Sisymbrium altissimum*
CRUCIFERAE H 20–100 cm. Erect, branched, annual herb. Lvs variable: middle lvs distinctly lobed; upper linear, stalkless. Fls <11 mm, with 4 yellow petals, in spikes on slender, spreading stalks, Jun–Aug. Frs pods, <10 cm. Waste places. Ge, Cz, Po, (but T, ex Fi). [**8**]

Garlic Mustard *Alliaria petiolata*
CRUCIFERAE H 20–120 cm. Erect, unbranched, rosette-forming, biennial herb with root smelling strongly of garlic. Lvs <10 cm, long-stalked, heart-shaped. Fls <6 mm, with 4 white petals, in terminal lfless clusters, Apr–Jun. Frs pods, <6 cm. Hedgerows. T, ex Ic. [**9**]

Thale Cress *Arabidopsis thaliana*
CRUCIFERAE H 5–50 cm. Erect, branched, annual herb, with stem hairy at top, smooth at bottom. Lvs variable: lower toothed, spoon-shaped, in rosette; upper narrowly oblong. Fls <3 mm, with 4 white petals, in short, lfless spikes, Apr–May. Frs slender pods, <18 mm. Walls, banks, waste places. T, ex Ic. [**10**]

Treacle Mustard *Erysimum cheiranthoides*
CRUCIFERAE H 15–90 cm. Erect, annual or overwintering herb, covered in branched, appressed hairs. Lvs variable: basal in rosette, oblong, dying before flowering; upper narrow, stalkless. Fls <6 mm, with 4 yellow petals, in terminal spike, Jun–Aug. Frs narrow pods, <25 mm. Arable fields, waste places. T. [**11**]

10
1
2
8
3
11
4
7
5
9
6

Winter-cress *Barbarea vulgaris* CRUCIFERAE H 30–90 cm. Erect, branching, hairless, biennial or perennial herb. Lvs variable: basal with 2–5 lateral lobes; upper ovate, simply toothed. Fls <9 mm, with 4 yellow petals, in dense, terminal cluster, May–Aug. Frs narrow pods, <25 mm, erect on short stalks. Hedgerows, meadows, stream-banks. Once valued as winter salad plant. T, (but Ic). [1]

Creeping Yellow-cress *Rorippa sylvestris* CRUCIFERAE H 20–50 cm. Creeping, hairless, perennial herb. Lvs divided into lanceolate, toothed segments. Fls <5 mm, with 4 yellow petals twice as long as sepals, in terminal clusters on wavy stems, Jun–Aug. Frs cylindrical pods, 6–18 mm, about as long as stalk. Wet areas, waste ground; often persists as weed on roadsides. T, (but Ic). [2]

Hairy Bitter-cress *Cardamine hirsuta* CRUCIFERAE H 7–30 cm. Erect, rosette-forming, annual herb with hairless stem, often branching from base. Lvs divided into 1–3 pairs of roundish lflets; terminal larger, kidney-shaped. Fls <7 mm, with 4 white petals, in terminal cluster, Apr–Aug. Frs erect, cylindrical pods, <25 mm, overtopping fls. Bare, dry places. T. [3] Wavy bittercress, *C. flexuosa*, very similar, but has 6 stamens.

Shepherd's-purse *Capsella bursa-pastoris* CRUCIFERAE H 3–40 cm. Erect, annual or biennial herb, hairless or with unbranched hairs. Lvs of basal rosette lanceolate, deeply lobed; upper less divided, clasping stem. Fls white, 2·5 mm across, all year. Frs pods, shaped like triangular purses. Arable land, waste ground. T. [4]

Field Penny-cress *Thlaspi arvense* CRUCIFERAE H 10–60 cm. Erect, hairless, annual herb, branched above, foetid when crushed. Lvs oblong or lanceolate, lower stalked, upper clasping stem. Fls <6 mm, with 4 white petals twice as long as sepals, in terminal cluster, May–Jul. Frs circular pods (pennies), <22 mm across, with deep, apical notch. Arable fields, waste places. T. [5]

Field Pepperwort *Lepidium campestre* CRUCIFERAE H 20–60 cm. Erect, branching, annual or biennial herb covered in dense, short, spreading hairs. Basal lvs stalked, fall before flowering; upper triangular, clasping stem. Fls <2·5 mm, with 4 white petals, in dense, terminal cluster, May–Aug. Frs round, winged pods, <4 mm across, with deep, apical notch, covered in scaly blisters. Waste ground. T, (but Ic). [6]

Hoary Cress *Cardaria draba* CRUCIFERAE H 30–90 cm. Erect, densely leafy, perennial herb with few hairs. Lvs oblong, clasping stem, slightly toothed. Fls <6 mm, with 4 white petals, in dense, flattened clusters, May–Jun. Frs heart-shaped pods, <4 mm across. Fields, roadsides; aggressive weed which has recently spread across Europe. (T, ex Ic, from S Europe.) [7]

Swine-cress *Coronopus squamatus* CRUCIFERAE H 5–30 cm. Prostrate, hairless, annual or biennial herb. Lvs divided into narrow segments. Fls <2·5 mm, with 4 white petals, arranged in short, inconspicuous spikes, Jun–Sep. Frs rounded, roughly ridged or warted pods, <4 mm across. Waste ground, esp where trampled. T, ex Ic. [8]

Charlock *Sinapis arvensis* CRUCIFERAE H 30–80 cm. Erect, often branched, annual herb, hairy at base. Lvs <20 cm, variable; lower divided, with large, terminal lobe; upper simple, lanceolate. Fls <2 cm, with 4 yellow petals, in terminal clusters, May–Jul. Frs pods, <4 cm, with long beak. Arable land. T, (but Ic). [9]

Weld *Reseda luteola* CRUCIFERAE H 50–150 cm. Erect, biennial herb with hollow stem. Lvs <12 cm, narrowly oblong, with wavy margins. Fls <5 mm, yellowish-green, with 4 lobed petals, in long, slender spikes, Jun–Aug. Frs 3-lobed capsules, <6 mm. Waste ground, open places. Once known as 'dyer's weed' because it produces brilliant yellow dye. Br, Ir, Fr, Lu, Be, De, Cz, Sw, (Po). [10]

1
2
3
4
5
6
7
8
9
10

Orpine *Sedum telephium* CRASSULACEAE H 20–60 cm. Erect, hairless, perennial herb with carrot-like, tuberous root. Lvs <8 cm, fleshy, ovate, with irregular teeth. Fls <12 mm, reddish-purple, with 5 petals, in flat-topped, terminal clusters, Jul–Sep. Woods, hedges. T, ex Ic. [1]

Meadow Saxifrage *Saxifraga granulata* SAXIFRAGACEAE H 10–50 cm. Erect, hairy, rosette-forming, perennial herb, over-wintering by bulbils in axils of lvs. Lvs <3 cm, kidney-shaped, with 7 or more shallow, rounded teeth, on long stalks. Fls <25 mm, white, with 5 petals, in loose clusters, Apr–Jun. Frs ovoid capsules, <8 mm. Grassland; becoming rare as old meadows ploughed. T, ex Ic. [2]

Rue-leaved Saxifrage *Saxifraga tridactylites* SAXIFRAGACEAE H 2–15 cm. Erect, hairy, usually branched, annual herb. Lvs variable, <1 cm: lower 3- to 5-lobed; upper entire, often tinged with red. Fls <5 mm, white, with 5 notched petals twice as long as sepals, Apr–Jun. Frs roundish capsules, <5 mm. Limestone rocks, walls. T, ex Ic. [3]

Red Currant *Ribes rubrum* GROSSULARIACEAE H 1–1·5 m. Erect, deciduous shrub. Lvs <10 cm, broader than long, hairy when young. Fls <5 mm, green, tinged purple, with 5 petals in drooping, elongated clusters, Apr–May. Frs red berries, <1 cm. Damp woods, hedges, often as escape from cultivation. Br, Fr, Lu, Be, Ne, Ge, (Ir, Cz, Po, Sw). [4] Black currant *R. nigrum* similar, but crushed lvs smell strongly; nearly always escape from cultivation. Br, Fr, Ge, Po, Fi, Sw, (Ir, Lu, Be, Ne, De, Cz, No).

Gooseberry *Ribes uva-crispa* GROSSULARIACEAE H 1–1·5 m. Much-branched, spiny, deciduous shrub. Lvs <5 cm, broader than long, deeply lobed, sometimes hairy. Fls <7 mm, with 5 pale pinkish-green sepals and white petals. Fls in clusters of 1–3, Mar–May. Frs green berries, <1 cm, usually hairy. Woods, hedges, often bird-sown and naturalized. Br, Fr, Lu, Be, Ne, Ge, Cz, Po, (Ir, De, Fi, Sw, No). [5]

Dropwort *Filipendula vulgaris* ROSACEAE H 15–80 cm. Erect, hairless, perennial herb with oval root-tubers. Basal lvs <25 cm, with 8–20 pairs of acutely lobed lflets; stem lvs few. Fls <2 cm, white, with 6 petals, in terminal clusters, May–Aug. Dry grasslands, esp where rich in lime. T, ex Ic. [6]

Lady's-mantle *Alchemilla vulgaris* ROSACEAE H 5–45 cm. Ascending or prostrate, perennial herb. Lvs <15 cm across, lobed into 7–11 broad-toothed segments, on long stalks. Fls <4 mm, green, no petals, in branched clusters, Jun–Sep. Very variable group of microspp (25 in N Europe). Damp grassland, rock ledges, woods. T. [7]

Parsley-piert *Aphanes arvensis* ROSACEAE H 2–20 cm. Inconspicuous, prostrate or ascending, hairy, annual herb, usually branched. Lvs <1 cm, fan-shaped, divided into 3–5 oblong lobes. Fls <2 mm, green, no petals, in tight clusters, half enclosed by green, lf-like, cup-shaped stipules, Apr–Oct. Arable fields, bare dry places. T, ex Ic, Fi, No. [8]

Dog Rose *Rosa canina* ROSACEAE H <5 m. Trailing, woody-stemmed, perennial herb with strong, curved prickles. Lvs divided into 2–3 pairs of oval lflets. Fls <6 cm, white or pink, with 5 petals, in small groups, on short stalks, Jun–Jul. Frs <2 cm, spherical, red when ripe. Woods, hedges, tracks. T, ex Ic. [9] Field rose *R. arvensis* similar, but has long, fused styles which persist on fr, T, ex SC. Downy rose *R. tomentosa* has straight prickles, densely hairy lvs. T, ex Ic.

Raspberry *Rubus idaeus* ROSACEAE H 100–150 cm. Erect, prickly, woody, biennial herb, suckering extensively. Lvs white below, divided into 3–5 ovate lflets <12 cm, terminal largest. Fls <1 cm, white, with 5 petals, in small, leafy clusters, Jun–Aug. Frs <25 mm, usually red. Woods, hedgerows. T, ex Ic. [10]

4
5
1
10
9
6
2
7
3
8

Bramble, Blackberry *Rubus fruticosus* ROSACEAE H <4 m. Woody perennial with biennial, trailing, prickly stems. Lvs 3- or 5-lobed. Fls white to deep pink, <3 cm across, in terminal clusters, May–Sep. Extremely variable: *c*2000 microspp recognized in Europe. Woods, thickets, hedges, waste places. T, ex Ic. [**1**]

Agrimony *Agrimonia eupatoria* ROSACEAE H 30–60 cm. Erect, hairy, perennial herb. Lvs divided into 3–6 pairs of oval, saw-toothed lflets <6 cm, with smaller lflets between. Fls <8 mm, yellow, with 5 petals, in terminal spike, Jun–Aug. Wood-margins, meadows. T, ex Ic. [**2**]

Salad Burnet *Sanguisorba minor* ROSACEAE H 15–60 cm. Erect, hairless, perennial herb. Lvs divided into 4–12 pairs of round, toothed lflets <2 cm, larger towards top. Fls <8 mm, green, tinged purple, no petals, in globular heads on long, lfless stems, May–Aug. Smells of cucumber when crushed. Grassland, esp calcareous. T, ex Ic, (but Fi, No.). [**3**]

Wood Avens *Geum urbanum* ROSACEAE H 20–60 cm. Erect, hairy, perennial herb. Lvs divided into 2–3 pairs of unequal, toothed lflets <1 cm, and single, large terminal one <8 cm. Fls <15 mm, yellow, with 5 petals, erect on long stalks, Jun–Aug. Woods. T, ex Ic. [**4**]

Wild Strawberry *Fragaria vesca* ROSACEAE H 5–30 cm. Perennial, rosette-forming herb with long, arching runners. Lvs divided into 3 toothed, ovate lflets <6 cm. Fls <18 mm, white, with 5 petals, Apr–Jul. Frs succulent, <2 cm. Wood margins, scrub. T. [**5**] Barren strawberry *Potentilla sterilis* similar but frs dry. T, ex Ic, Fi, No.

Tormentil *Potentilla erecta* ROSACEAE H 10–30 cm. Creeping, perennial herb. Lvs divided into 3–5 toothed, ovate to lanceolate lflets <3 cm. Fls <11 mm, yellow, with 4 petals, Jun–Sep. Fens, bogs, heaths, non-calcareous grassland. T. [**6**] Creeping cinquefoil *P. reptans* similar, but fls with 5 petals; hedge-banks, waste places, rarely on acid soils. T, ex Ic.

Hawthorn *Crataegus monogyna* ROSACEAE H <10 m. Prickly, woody shrub or tree. Lvs <3·5 cm, obovate to rhomboid, 3- to 7-lobed. Fls single or double, white or pink, <15 mm across, May–Jun. Frs red berries. Hedgerows, scrub. T, ex Ic. [7]

Crab Apple *Malus sylvestris* ROSACEAE H 2–10 m. Deciduous tree with dense, round crown. Lvs <10 cm, ovate to almost round, hairless when mature. Fls white or pink, with 5 petals, in umbel-like clusters, May. Frs <3 cm, yellowish-green. Woods, hedges. T, ex Ic. [8]

Common Whitebeam *Sorbus aria* ROSACEAE H <25 m. Deciduous tree or shrub with wide crown. Lvs <12 cm, ovate to elliptical, toothed, densely white, hairy beneath. Fls <15 mm, white, with 5 petals, in flat-topped cluster, May–Jun. Frs <15 mm, longer than broad, scarlet. Woods, scrub, on calcareous soils. Br, Ir, Fr, Lu, Be, Ge, Cz, Po. [9]

Blackthorn *Prunus spinosa* ROSACEAE H <4 m. Much-branched, thorny, deciduous shrub, spreading by suckers, forming dense thickets. Lvs <4 cm, elliptical, with rounded teeth. Fls <15 mm, white, with 5 petals; before lvs, Mar–May. Frs bluish, sour sloes. Woods, scrub, hedges. T, ex Ic. [10]

Wild Cherry *Prunus avium* ROSACEAE H <25 m. Deciduous tree, suckering freely, with bark peeling in thin strips. Lvs <15 cm, oblong, with rounded teeth. Fls <25 mm, white, with 5 petals, in umbels of 2–6, Apr–May. Frs <1 cm, red, sweet or bitter. Woods, hedges. T, ex Ic, Fi. [11]

Bird Cherry *Prunus padus* ROSACEAE H <15 m. Deciduous tree or shrub with ascending branches, strong-smelling bark. Lvs <10 cm, elliptical, finely toothed. Fls <15 mm, white, with 5 petals, in spikes of 15–35, May. Frs <8 mm, black, bitter. Woods. T, ex Ic. [12]

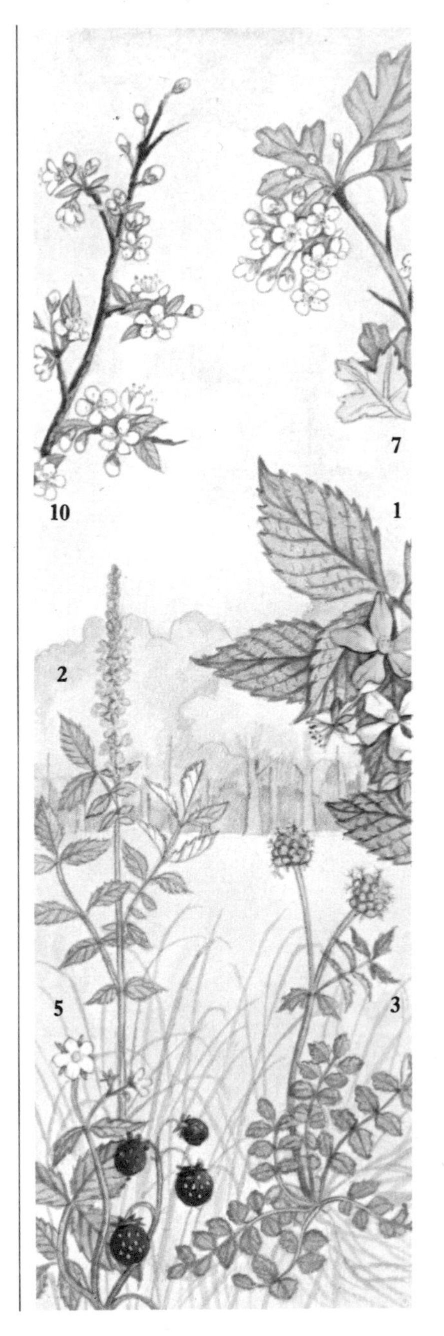

Wild Liquorice *Astragalus glycyphyllos* LEGUMINOSAE H 30–100 cm. Prostrate, perennial herb. Lvs <20 cm, divided into 4–6 pairs of ovate lflets <4 cm. Fls <15 mm long, pale greenish-yellow, in long-stalked spikes, Jul–Aug. Frs slightly curved pods, <35 mm. Rough grass, bushy places. T, ex Ic, (but Fi). [1]

Hairy Tare *Vicia hirsuta* LEGUMINOSAE H 20–70 cm. Trailing, annual herb. Lvs <5 cm, divided into 4–8 pairs of alternate, linear, notched lflets <12 mm. Fls <5 mm long, dirty white or purple, in long-stalked clusters of 1–9, May–Aug. Frs hairy pods with 2 seeds, <1 cm. Hedges, railway-banks. T, (but Ic). [2]

Tufted Vetch *Vicia cracca* LEGUMINOSAE H 60–200 cm. Hairy, scrambling, perennial herb. Lvs <10 cm, divided into 6–12 pairs of linear-lanceolate lflets <25 mm. Fls <12 mm long, bluish-violet, in long-stalked spikes of 10–40, Jun–Aug. Frs pods with 2–6 seeds, <20 mm. Hedgerows, grassland. T. [3]

Bush Vetch *Vicia sepium* LEGUMINOSAE H 30–100 cm. Hairy, scrambling, perennial herb. Lvs <10 cm, divided into 5–9 pairs of elliptical, notched lflets <3 cm. Fls <15 mm long, pale purple, in short-stalked clusters of 2–6, May–Aug. Frs beaked, hairless pods with 6–10 seeds, <25 mm. Hedgerows, roadsides, thickets. T. [4]

Common Vetch *Vicia sativa* LEGUMINOSAE H 15–100 cm. Hairy, trailing or climbing, annual herb. Lvs <8 cm, divided into 4–8 pairs of linear to ovate lflets <2 cm; stipules at base of each lf with dark blotch. Fls <3 cm, purple, solitary or in pairs, May–Sep. Frs almost hairless pods with 4–12 seeds, <80 mm. Hedges, grassland, arable. T, (but Ic). [5]

Meadow Vetchling *Lathyrus pratensis* LEGUMINOSAE H 30–120 cm. Scrambling, finely hairy, perennial herb. Lvs <5 cm, with 1 pair of lanceolate lflets <3 cm, arrowhead-shaped stipules. Fls <18 mm long, yellow, in long-stalked clusters of

5–12, May–Aug. Frs pods, 5–10 seeds, <35 mm. Hedgerows, grassland. T. [6]

Narrow-leaved Everlasting-pea *Lathyrus sylvestris* LEGUMINOSAE H 60–200 cm. Scrambling, perennial herb. Lvs divided into 1 pair of linear to lanceolate lflets <15 cm. Fls <2 cm, rose-pink, in clusters of 3–8 on stalk <20 cm. Frs smooth, brown pods with 8–14 seeds, <7 cm. Woods, scrub. T, ex Ir, Ic.

Bitter Vetch *Lathyrus montanus* LEGUMINOSAE H 15–40 cm. Erect, perennial herb with winged stems. Lvs divided into 2–4 pairs of narrow to elliptical lflts <4 cm. Fls crimson, turning blue or green, <12 mm long, in clusters of 2–4, Apr–Jul. Frs smooth, reddish-brown pods, <4 cm. Woods, shady hedge-banks. T, ex Ic.

Spring Pea *Lathyrus vernus* LEGUMINOSAE H 20–40 cm. Scrambling, hairless, perennial herb. Lvs <10 cm, no tendrils, with 2–4 pairs of lanceolate to ovate, soft, shining lflets <6 cm. Fls <2 cm long, purple, turning bluish-purple, in long-stalked, loose clusters of 3–10, Apr–Jun. Frs hairless pods with 8–14 seeds, <6 cm. Woods, thickets. Fr, De, Ge, Cz, Po, Fi, Sw, No, (Be, Ne). [7]

Spiny Restharrow *Ononis spinosa* LEGUMINOSAE H 30–60 cm. Erect, branched, perennial herb; spiny stems. Lvs with 3 narrowly elliptical, toothed lflets <2 cm. Fls <2 cm long, pink, solitary or in short-stalked clusters of 2–3, Jun–Sep. Frs few-seeded pods, <1 cm. Rough grassland. T, ex Ir, Ic, Fi. [8]

Common Restharrow *Ononis repens* LEGUMINOSAE H 30–60 cm. Shrubby, perennial herb with creeping or ascending branches, stems hairy lacking spines. Lvs with 3 ovate, blunt or notched lflets <2 cm. Fls <2 cm long, pink or purple, in loose, lfy spikes, Jun–Sep. Frs pods with 1–4 seeds, <7 mm. Rough, calcareous grassland. T, ex Ic, (but Fi).

Ribbed Melilot *Melilotus officinalis* LEGUMINOSAE H 60–120 cm. Erect, branched, hairless, biennial herb. Lvs of 3 narrowly elliptical, toothed lflets <2 cm. Fls <6 mm long, bottom petal shorter than others, yellow, in long spikes, Jul–Sep. Frs ribbed, hairless pods, <5 mm. Waste places. T, ex Ic, (but not native in north). [1] White melilot *M. alba*, similar in distribution and appearance, has white fls.

Black Medick *Medicago lupulina* LEGUMINOSAE H 5–50 cm. Procumbent or ascending, hairy, annual or short-lived perennial herb. Lvs of 3 rounded, notched lflets <2 cm. Fls <3 mm long, bright yellow, in compact, stalked heads of 10–15, Apr–Aug. Frs kidney-shaped pods, <2 mm, black when ripe. Grassy places, roadsides, esp where calcareous. T, (but Ic). [2]

Hop Trefoil *Trifolium campestre* LEGUMINOSAE H 10–30 cm. Erect or ascending, hairy, annual herb. Lvs of 3 oval lflets <1 cm, broadest at top. Fls <5 mm long, yellow, in compact, short-stalked heads of *c*40, Jun–Sep. Frs ovoid, 1-seeded pods, <2·5 mm. Dry grassland, roadsides, esp where calcareous. T, ex Ic, (but Fi.) [3] Lesser trefoil *T. dubium*, has heads of 10–20 fls <3 mm long.

Hare's-foot Clover *Trifolium arvense* LEGUMINOSAE H 5–20 cm. Erect or ascending, softly hairy, branched, annual herb. Lvs of 3 oblong lflets <2 cm. Fls <3·5 mm long, white or pink, in cylindrical, short-stalked heads <25 mm, Jun–Sep. Frs pods, <1·3 mm. Dry, sandy places. T, ex Ic. [4]

Red Clover *Trifolium pratense* LEGUMINOSAE H <60 cm. Spreading or erect, somewhat hairy, perennial herb. Lvs of 3 elliptical lflets <3 cm, with white, crescent-shaped spot near base. Fls purplish-pink, in globular heads, <3 cm, May–Sep. Meadows, pastures, grassland of all kinds; widely cultivated as forage crop. T. [5]

Zigzag Clover *Trifolium medium* LEGUMINOSAE H 30–50 cm. Straggling, hairy, perennial herb with branched, zigzag stems and creeping underground stems. Lvs of 3 narrowly elliptical lflets <4 cm. Fls <2 cm long, deep rose-pink, in round, somewhat flattened heads <3·5 cm, on long stalks, Jun–Sep. Frs pods, <2 mm. Open woods, grassland. T, ex Ic. [6]

Alsike Clover *Trifolium hybridum* LEGUMINOSAE H 20–40 cm. Erect or ascending, perennial herb. Lvs of 3 ovate lflets <2 cm, broadest near top. Fls <1 cm long, purple or white-turning-pink, in spherical heads <25 mm, on long stalks, Jun–Sep. Frs pods, <4 mm. Grassland, roadsides, always escape from cultivation as forage crop. (T, ex Ic). [7]

Bird's-foot Trefoil *Lotus corniculatus* LEGUMINOSAE H 10–40 cm. Trailing, sometimes hairy, perennial herb. Lvs of 5 oval lflets <1 cm, on short stalks. Fls 15 mm long, yellow tipped with red, in flattened clusters of 2–6, on long stalks, Jun–Sep. Frs pods, <3 cm. Grassland, roadside verges. T, (but Ic). [8]

Horseshoe Vetch *Hippocrepis comosa* LEGUMINOSAE H 10–40 cm. Almost hairless, spreading, perennial herb. Lvs of 3–8 pairs of ovate to linear lflets <15 mm. Fls <1 cm long, yellow, in flattened clusters of 5–12, on stalks longer than lvs, May–Jul. Frs flattened pods, <3 cm, breaking up into horseshoe-shaped segments. Dry grassland, esp where calcareous. Br, Fr, Lu, Be, Ne, Ge, Cz. [9]

Wood-sorrel *Oxalis acetosella* OXALIDACEAE H 5–15 cm. Slightly hairy, creeping, perennial herb with underground stem clothed in swollen lf-bases. Lvs of 3 heart-shaped lflets <2 cm, broadest above, yellow-green. Fls <15 mm, of 5 white petals veined with lilac, solitary, on long stalks, Apr–May. Frs ovoid capsules, <4 mm. Woods, shady places. Lvs, which may be eaten as salad, fold up in bright sunlight and at night. T. [10]

1
2
3
4
5
6
7
8
9
10

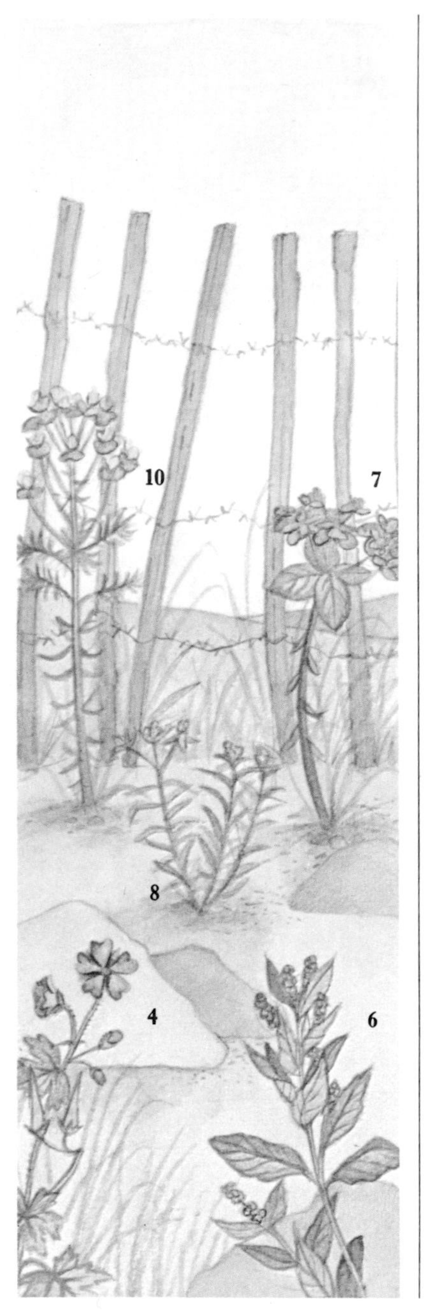

Hedgerow Crane's-bill *Geranium pyrenaicum* GERANIACEAE H 25–60 cm. Erect, glandular-hairy, perennial herb. Lvs round, <8 cm across, divided into 5–9 lobes, each often further 3-lobed or toothed. Fls <15 mm, in pairs, with 5 purple, deeply notched petals, Jun–Aug. Frs with upright beak <1 cm, and 5 1-seeded lobes which fling out seed as they spring upwards when ripe. Waysides, hedgerows. (T, ex Ic, from S Europe). [1]

Meadow Crane's-bill *Geranium pratense* GERANIACEAE H 30–80 cm. Erect, densely glandular-hairy, perennial herb. Lvs polygonal, <15 cm across, divided into 5–7 lobes, each further divided into 3–4 toothed lobes; turning red in autumn. Fls <3 cm, in pairs, with 5 violet-blue, beautifully veined petals, Jun–Sep. Frs with upright beak, <2·5 cm. Meadows, roadsides. T, ex Ic, (but De). [2]

Cut-leaved Crane's-bill *Geranium dissectum* GERANIACEAE H 10–60 cm. Straggling, hairy, annual or biennial herb. Lvs circular, <7 cm across, divided almost to centre into 5–7 narrow lobes, each further divided into 3–5 toothed lobes. Fls <1 cm, in pairs, with 5 reddish-pink, notched petals, on stalks <15 mm, May–Aug. Frs with beak <12 mm. Fields, hedges, walls. T, ex Ic. [3]

Dove's-foot Crane's-bill *Geranium molle* GERANIACEAE H 10–40 cm. Straggling, glandular-hairy, annual herb. Lvs circular, <5 cm across, divided into 5–9 lobes twice as long as undivided portion, each further 3-lobed at apex. Fls <15 mm, in pairs, with 5 rose-purple petals, Apr–Sep. Frs with beak <8 mm. Dry grassland, waste places. T, ex Ic. [4]

Dog's Mercury *Mercurialis perennis* EUPHORBIACEAE H 15–40 cm. Erect, hairy, perennial herb with long, creeping, underground stems, forming large patches. Lvs opposite, <8 cm, elliptical to ovate, toothed, on short stalks <1 cm. Fls <5 mm, green with 3 sepals, ♂ and ♀ on separate plants, ♂ fls in spike-like clusters,

♀ 1–3 on long stalks, Feb–Apr. Woods. Poisonous. T, ex Ic, (but Ir). [♂ 5]

Annual Mercury *Mercurialis annua* EUPHORBIACEAE H 10–50 cm. Erect, hairless, yellow-green, annual herb, branching from base. Lvs opposite, <5 cm, elliptical to ovate, toothed, on stalks <15 mm. Fls <5 mm, green, with 3 sepals, ♂ and ♀ on separate plants, ♂ fls in erect spike <5 mm, ♀ almost stalkless, Jul–Oct. Arable fields, waste places. T, ex Ic, (but Ir, De, FS). [♀ 6]

Sun Spurge *Euphorbia helioscopia* EUPHORBIACEAE H 10–50 cm. Erect, hairless, yellow-green, annual herb. Lvs <3 cm, with toothed, blunt tips, tapering to narrow base. Fls in flat-topped, golden-yellow, 5-rayed, compound umbels, the ultimate branches with 2–3 'fls', each of one 3-styled ♀ fl and several 1-anthered ♂ fls lacking sepals or petals but surrounded by petal-like bracts, May–Oct. Frs smooth capsules, <3 mm, dividing into 3. Arable fields. T, ex Ic. [7]

Dwarf Spurge *Euphorbia exigua* EUPHORBIACEAE H 5–30 cm. Similar to *E. helioscopia*, but with linear, usually pointed lvs, <3 cm; inflor of 3-rayed umbels, Jun–Oct; frs rough capsules, <2 mm. Arable fields. T, ex Ic, Fi. [8]

Wood Spurge *Euphorbia amygdaloides* EUPHORBIACEAE H 30–80 cm. Erect, hairy, perennial herb. Lvs <8 cm, lanceolate, blunt, broadest near top, dark green. Fls set in yellowish, kidney-shaped bracts in 5- to 10-rayed umbels, Mar–May. Frs glandular capsules, <4 mm. Woods. Br, Ir, Fr, Lu, Be, Ne, Ge, Cz, Po. [9]

Cypress Spurge *Euphorbia cyparissias* EUPHORBIACEAE H 10–30 cm. Erect, hairless, perennial herb with long, creeping underground stem. Lvs <3 cm, narrow, greenish-yellow. Fls set in yellowish, kidney-shaped bracts, often turning reddish, in 9- to 15-rayed umbels, May–Aug. Frs roughened capsules, <3 mm. Cultivated ground. Fr, Lu, Be, Ne, Ge, Cz, Po, (Br, De, FS). [10]

Field Maple *Acer campestre* ACERACEAE H <15 m. Small, deciduous tree or shrub. Lvs opposite, <7 cm, divided into 3–5 triangular lobes. Fls <6 mm, of 5 pale green petals, in erect spikes of 10–20, appearing with lvs, May–Jun. Frs paired, winged. Woods, scrub. T, ex Ic, Fi, No, (but Ir). [1]

Sycamore *Acer pseudoplatanus* ACERACEAE H <30 m. Large, deciduous tree with spreading crown. Lvs opposite, <16 cm, divided into 5 triangular lobes. Fls <6 mm, of 5 yellowish-green petals, in drooping spikes of 60–100, appearing with or after lvs, Apr–Jun. Frs paired, winged. Woods, hedges. Fr, Lu, Be, Ne, Ge, Cz, Po, (Br, Ir, De, Sw). [2]

Holly *Ilex aquifolium* AQUIFOLIACEAE H <10 m. Evergreen tree or shrub, with familiar prickly, dark green leaves. Fls white, small, ♂ and ♀ on different trees. ♀ trees produce bright red berries, late autumn–winter. Woods, scrub, hedgerows. T, ex Ic, Cz, Po, Fi, Sw.

Spindle *Euonymus europaeus* CELASTRACEAE H 2–6 m. Much-branched, hairless, deciduous shrub or small tree with 4-angled twigs. Lvs opposite, <13 cm, ovate-lanceolate. Fls <1 cm, of 4 greenish petals, in axillary clusters of 3–10, May–Jun. Frs deep pink, 4-lobed capsules, <15 mm across, when open exposing bright orange flesh which surrounds seed. Woods, hedges. Wood used for skewers. T, ex Ic, Fi. [3]

Buckthorn *Rhamnus catharticus* RHAMNACEAE H 4–6 m. Thorny, hairless, deciduous shrub. Lvs opposite, <6 mm, elliptical, toothed, with 2–3 pairs of conspicuous veins. Fls <4 mm, of 4 greenish petals, solitary or in small axillary clusters, May–Jun. Frs berries, black when ripe, <1 cm, poisonous. Woods, scrub, hedges. T, ex Ic. [4]

Small-leaved Lime *Tilia cordata* TILIACEAE H <30 m. Spreading, deciduous tree. Lvs <6 cm, heart-shaped at base, toothed. Fls <1 cm, of 5 yellowish-white petals, in clusters of 4–10, Jul. Frs rounded, <6 mm, on winged bract. Woods. T, ex Ic, Ir. [5]

Musk Mallow *Malva moschata* MALVACEAE H 30–80 cm. Erect, hairy, perennial herb, basally branched, with stems often purple-spotted. Lvs circular, <8 cm across, much-divided into linear segments. Fls <6 cm, of 5 rose-pink (rarely white), notched petals, Jul–Aug. Frs blackish when ripe. Grassland, wood margins. T, ex Ic, (but De, FS). [6]

Common Mallow *Malva sylvestris* MALVACEAE H 45–90 cm. Erect or ascending, sparsely hairy, perennial herb. Lvs roundish, <10 cm across, divided into 5–7 rather deep, coarsely toothed lobes. Fls <4 cm, of 5 rose-purple petals, Jun–Sep. Frs brownish-green when ripe. Roadsides, waste places. T, ex Ic. [7]

Dwarf Mallow *Malva neglecta* MALVACEAE H 15–60 cm. Trailing or ascending, hairy, annual or longer-lived herb. Lvs roundish, <7 cm across, divided into 5–7 shallow, irregularly toothed lobes. Fls <25 mm, of 5 whitish, lilac-veined petals, Jun–Sep. Frs brownish-green. Roadsides, waste places. T, ex Ic, (but Fi). [8]

Hairy St John's-wort *Hypericum hirsutum* HYPERICACEAE H 40–100 cm. Erect, hairy, perennial herb. Lvs opposite, <5 cm, ovate, blunt, stalkless, with translucent glands on blade. Fls <15 mm, of 5 pale yellow petals, in loose, many-fld, cylindrical spikes, Jul–Aug. Frs dry capsules. Woods, damp grassland, shady places. T, ex Ic. [9]

Perforate St John's-wort *Hypericum perforatum* HYPERICACEAE H 30–90 cm. Erect, hairless, perennial herb; stems woody at base, with 2 raised lines. Lvs opposite, <2 cm, elliptical, stalkless, with translucent glands on blade. Fls <2 cm, of 5 golden-yellow petals, in many-fld, branched clusters, Jun–Sep. Frs dry capsules. Open woods, dry grassland. T, ex Ic. [10]

1
2
3
4
5
6
7
8
9
10

Sweet Violet *Viola odorata* VIOLACEAE H 5–15 cm. Hairy, sweet-scented, perennial herb. Lvs <6 cm, heart-shaped, bluntly toothed, on long stalks with down-turned hairs. Fls <15 mm, of 5 deep violet or white petals, with thick spur, solitary, Feb–May. Frs hairy, spherical capsules, <1 cm. Woods, thickets, hedgerows. T, ex Ic, Fi. [1]

Hairy Violet *Viola hirta* VIOLACEAE H 5–15 cm. Hairy, rosette-forming, scentless, perennial herb. Lvs <6 cm, narrowly heart-shaped, on long stalks with numerous spreading hairs. Fls <15 mm, of 5 bright blue-violet petals, with darker spur, solitary, Apr–May. Frs hairy, spherical capsules, <1 cm. Calcareous pastures, woods. T, ex Ic, (but Fi). [2]

Common Dog-violet *Viola riviniana* VIOLACEAE H 2–20 cm. Almost hairless, rosette-forming, perennial herb. Lvs <8 cm, heart-shaped, bluntly toothed, on long, hairless stalks. Fls <2 cm, of 5 blue-violet petals, with thick, paler spur notched at end, solitary, Apr–Jun. Frs hairless, 3-angled capsules, <13 mm. Woods, hedgerows, heaths. T. [3]

Field Pansy *Viola arvensis* VIOLACEAE H 10–40 cm. Very variable, erect, branched, somewhat hairy, annual herb. Lvs <3 cm, lower roundish, upper lanceolate to elliptical and bluntly toothed. Fls <2 cm from top to bottom, of 5 petals hardly longer than sepals, cream or with some blue-violet, solitary, Apr–Oct. Frs globular capsules, <1 cm. Arable fields, waste places. T, ex Ic. [4]

Common Rock-rose *Helianthemum chamaecistus* CISTACEAE H 5–30 cm. Low-growing, perennial herb. Lvs <2 cm, oblong, green above, white and hairy below. Fls yellow, with 5 petals, *c*25 mm across, Jul–Sep. Base-rich grassland. T, ex Ic, No. [5]

Enchanter's-nightshade *Circaea lutetiana* ONAGRACEAE H 20–70 cm. Erect or ascending, somewhat glandular-hairy, perennial herb. Lvs opposite, <10 cm, ovate with heart-shaped base, sparsely toothed. Fls <8 mm, of 2-notched, white or very pale pink petals, in dense clusters, Jun–Aug. Frs 2-seeded, covered in hooked bristles. Woods. T, ex Ic, Fi. [6]

Rosebay Willowherb *Epilobium angustifolium* ONAGRACEAE H 30–120 cm. Robust, perennial herb. Lvs alternate, long, narrow, 5–15 cm. Fls pink, 2–3 cm across, in long, dense spikes, Jul–Sep. Frs capsules, <8 cm, splitting to release white-plumed, wind-borne seeds. Woodland clearings, waste places. T. [7]

Broad-leaved Willowherb *Epilobium montanum* ONAGRACEAE H 20–60 cm. Erect, hairy, perennial herb. Lvs opposite, <7 cm, ovate. Fls <9 mm, of 4 notched, pink petals, in long, lfy spikes, Jun–Aug. Frs capsules, <8 cm, opening to release fluffy seeds. Woods, hedges, waste places. T, ex Ic. [8]

American Willowherb *Epilobium adenocaulon* ONAGRACEAE H 60–90 cm. Erect, glandular-hairy, perennial herb with reddish stems. Lvs opposite, <10 cm, lanceolate. Fls <6 mm, of 4 notched, pale pink petals, in long, lfy spikes, Jun–Aug. Frs capsules, <65 mm, covered in glandular hairs. Waste places. (T, ex Ir, Ic, from N America.) [9]

Dogwood *Cornus sanguinea* CORNACEAE H <4 m. Deciduous shrub with purplish-red stems. Lvs opposite, <8 cm, ovate, on grooved stalks <15 mm. Fls <15 mm, of 4 dull white petals, in many-fld, flat-topped clusters <5 cm across, Jun–Jul. Frs black, globular berries, <8 mm. Scrub, hedgerows, calcareous soils. T, ex Ic, (but Fi). [10]

Ivy *Hedera helix* ARALIACEAE H <30 m. Woody evergreen. Lvs of non-flowering shoots 3- to 5-lobed, of flowering shoots elliptical. Fls yellowish-green, <1 cm across, arranged in umbels, Sep–Nov. Frs black berries, with 2–3 seeds. On trees in woods, hedgerows. T, ex Ic, Fi. [11]

11
6
8
10
9
7
5
2
4
3
1

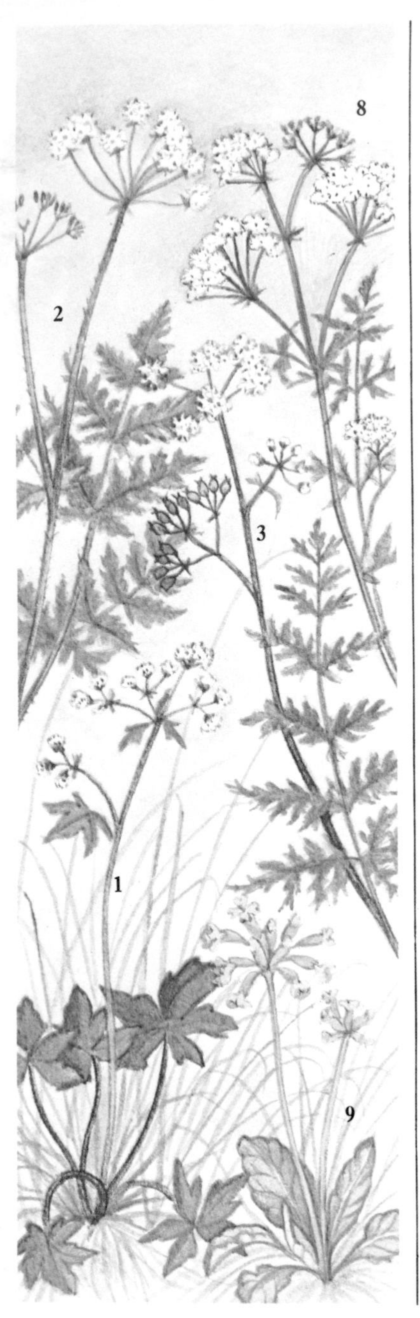

Sanicle *Sanicula europaea* UMBELLIFERAE H 20–60 cm. Erect, hairless, perennial herb. Lvs <6 cm, round, divided into 3–5 toothed lobes, on long stalks <25 cm. Fls <2 mm, of 5 notched, pink or white petals, in primary umbels of *c*4 rays, May–Sep. Frs ovoid, <3 mm, with numerous hooked bristles, dispersed by animals. Woods. T, ex Ic. [**1**]

Rough Chervil *Chaerophyllum temulentum* UMBELLIFERAE H 30–100 cm. Erect, hairy, perennial herb with solid, purple-spotted stems. Lvs <20 cm, 2–3 times divided into coarsely toothed lobes. Fls <2 mm, of 5 notched, white petals, in umbels <6 cm across of 8–10 rays, Jun–Jul. Frs oblong-ovoid, <7 mm, often purple. Open woods, hedge-banks. T, ex Ic, Fi, No. [**2**]

Cow Parsley *Anthriscus sylvestris* UMBELLIFERAE H 60–100 cm. Erect, downy, biennial herb with hollow, furrowed stems. Lvs <30 cm, 2–3 times divided into coarsely toothed lobes. Fls <4 mm, of 5 notched, white petals, in umbels <6 cm across of 4–15 rays, Apr–Jun. Frs oblong-ovoid, <5 mm. Open woods, hedgerows. T, ex Ic. [**3**]

Burnet Saxifrage *Pimpinella saxifraga* UMBELLIFERAE H 30–100 cm. Slender, erect, perennial herb. Lvs variable, <10 cm, lower once-divided into 3–7 pairs of ovate, toothed segments <25 mm, upper 1–2 times divided into linear segments. Fls <2 mm, of 5 white petals in umbels <6 cm across of 10–20 rays which droop in bud, Jun–Jul. Frs broadly ovoid, <2·5 mm. Dry grassland. T, ex Ic, Fi. [**4**]

Ground-elder *Aegopodium podagraria* UMBELLIFERAE H 40–100 cm. Hairless, low-growing, perennial herb with far-creeping underground stems. Lvs divided into 3 lflets which may be further divided into 3. Fls white, in umbels 2–6 cm across of 15–20 rays, May–Jul. Frs oval, 4 mm. Woods, hedges, waste places, gardens. T, (but Br, Ir, Ic). [**5**]

Hemlock *Conium maculatum* UMBELLIFERAE H 50–250 cm. Erect, branched, hairless, biennial herb with purple-spotted, hollow, furrowed stems. Lvs <30 cm, 2–4 times divided. Fls <2 mm, of 5 white petals in umbels <5 cm across of 10–20 rays, Jun–Jul. Frs rounded with wavy, toothed ridges, <3 mm. Damp hedgerows, waste places. Extremely poisonous. T, ex Ic. [6]

Hogweed *Heracleum sphondylium* UMBELLIFERAE H 50–200 cm. Erect, hairy, branched, biennial herb with hollow stems. Lvs variable, <60 cm, once-divided into shallow lobes or linear segments. Fls <1 cm, of 5 deeply notched, white or pink petals, outer often larger, in umbels <15 cm across of 7–20 rays, Jun–Sep. Frs rounded, whitish, <1 cm. Woodland margins, hedgerows. T, ex Ic. [7]

Upright Hedge-parsley *Torilis japonica* UMBELLIFERAE H 30–125 cm. Erect, hairy, annual or biennial herb with solid stems. Lvs <20 cm, 1–3 times divided into ovate to lanceolate, toothed segments. Fls <3 mm, of 5 pink or purplish-white petals, outer often larger, in umbels <4 cm across of 5–12 rays, Jul–Aug. Frs covered in hooked spines, <4 mm. Wood margins, hedges. T, ex Ic. [8]

Cowslip *Primula veris* PRIMULACEAE H 10–30 cm. Rosette-forming, hairy, perennial herb. Lvs <20 cm, ovate to oblong, with blunt teeth and winged stalks. Fls <15 mm, of 5 joined, yellow petals with orange spot at base, in umbels of 3–30 on erect, lfless stalks, Apr–May. Frs ovoid capsules, <1 cm, enclosed by calyx. Open woods, meadows. T, ex Ic. [9]

Primrose *Primula vulgaris* PRIMULACEAE H 5–20 cm. Rosette-forming, perennial herb. Lvs <25 mm, ovate-oblong. Fls <4 cm, of 5 joined, yellow (rarely pink) petals, solitary, on long stalks <10 cm, Feb–May. Frs ovoid capsules, enclosed by calyx. Woods, hedge-banks. Br, Ir, Fr, Be, Ne, De, Ge, Cz, No. [10]

Ash *Fraxinus excelsior* OLEACEAE H <40 m. Deciduous tree, with smooth, grey bark when young, and black buds. Lvs opposite, <30 cm, pinnate with <13 lflets. Fls purplish, without petals or sepals, appearing before lvs, Apr–May. Frs winged 'keys', <5 cm. Woods, hedgerows, on wide range of dry and moist soils. Springy wood widely used for tool handles. T, ex Ic, Fi. [1]

Privet *Ligustrum vulgare* OLEACEAE H <5 m. Partially evergreen shrub. Lvs opposite, 3–6 cm, lanceolate. Fls white, short-stalked, 4–5 mm, fragrant. Frs black berries, 6–8 mm. Woodland margins, hedgerows, scrub, esp on calcareous soils. So named because used in hedges to make 'privy' (private) gardens. T, ex Ic, De. [2]

Common Centaury *Centaurium erythraea* GENTIANACEAE H 10–50 cm. Slender, erect, hairless, rosette-forming, annual herb, branched above. Lvs opposite, <5 cm, ovate or elliptical, with blunt tips. Fls <12 mm, of 5 joined, pink-purple petals, in dense, terminal clusters, Jun–Oct. Frs cylindrical capsules, longer than calyx. Dry grassland, open woods. T, ex Ic, Fi, No. [3]

Autumn Gentian *Gentianella amarella* GENTIANACEAE H 5–30 cm. Slender, erect, rosette-forming, biennial herb. Basal lvs ovate to tongue-shaped, upper lanceolate. Fls <22 mm, of 4–5 equal sepals and 4–5 joined, blue-purple, pink or white petals, terminal on stems and branches, Jul–Oct. Frs stalkless capsules. Short grass, dunes. T, ex Lu, Ne. [4] Field gentian *G. campestris*, of more northerly distribution, has 2 pairs of unequal sepals and 4 petals.

Field Madder *Sherardia arvensis* RUBIACEAE H 5–40 cm. Erect or spreading, hairless, annual herb with rough, 4-angled stems. Lvs <18 mm, ovate, toothed, with acute points, 4–6 in whorl. Fls <3 mm, of 4 joined, pale lilac petals, in terminal heads of 4–8, May–Oct. Frs 2-celled, covered with short bristles. Arable fields, waysides. T, ex Ic. [5]

Hedge Bedstraw *Galium mollugo* RUBIACEAE H 25–120 cm. Erect or straggling, hairless, perennial herb with 4-angled stems swollen at nodes. Lvs <25 mm, linear, toothed, with acute points, 6–8 in whorl. Fls <3 mm, of 4 joined, white petals, in loose, terminal clusters, Jun–Sep. Frs 2-celled, rough, hairless, <1 mm. Meadows, roadsides, on calcareous soils. T, ex Ic. [6]

Woodruff *Galium odoratum* RUBIACEAE H 15–45 cm. Creeping, perennial herb with erect, 4-angled stems. Lvs <4 cm, lanceolate or elliptical, toothed, 6–8 in whorl. Fls <6 mm, of 4 joined, white petals, in umbel-like clusters, May–Jun. Frs 2-celled, rough, with hooked bristles, <3 mm. Woods. Hay-scented when dry and once hung in garlands in churches. T, ex Ic. [7]

Crosswort *Cruciata laevipes* RUBIACEAE H 15–70 cm. Creeping, much-branched, hairy, perennial herb with slender, 4-angled, ascending stems. Lvs <25 mm, ovate-elliptical, 4 in whorl, yellowish-green. Fls <2·5 mm, of 4 joined, pale yellow petals, in clusters of *c*8 in axils of lvs, May–Jun. Frs spherical, <1·5 mm. Open woods, scrub, hedgerows, esp on calcareous soils. Br, Fr, Lu, Be, Ne, Ge, Cz, Po, (Ir). [8]

Field Bindweed *Convolvulus arvensis* CONVOLVULACEAE H 20–200 cm. Scrambling or twining, perennial herb with long, creeping underground stems. Lvs <5 cm, arrow-shaped, with spreading basal lobes. Fls <2 cm, funnel-shaped, pink and/or white, usually solitary, long-stalked, Jun–Sep. Frs 2-celled capsules, <3 mm. Cultivated ground, waste places. Extremely resistant to drought and weedkillers. T, ex Ic. [9] Hedge bindweed *Calystegia sepium*, similar in habitat and distribution, has lvs <15 mm and fls <7 cm with sepals hidden by inflated, heart-shaped, lfy bracts.

1
2
3
4
5
6
7
8
9

Russian Comfrey *Symphytum × uplandicum* BORAGINACEAE H 50–200 cm. Roughly hairy, perennial herb with thick underground stems. Lvs <25 cm, lanceolate, hardly running down stem. Fls <18 mm, pink, purple or violet, tubular, in terminal clusters, May–Oct. Frs shining black nutlets. Widely cultivated, escaping to roadsides, ditch banks, waste places. (Br, Ir, Fr, Lu, Be, Ne, De, Fi, Sw, No). [1]

Lungwort *Pulmonaria officinalis* BORAGINACEAE H 10–30 cm. Hairy, perennial herb with creeping underground stems. Lvs <10 cm, broadly ovate, white-spotted, lower narrowed into winged stalk, upper often half-clasping stem. Fls <1 cm across, funnel-shaped, pink then blue, in terminal clusters, Mar–Jun. Frs pointed nutlets, <4 mm. Damp woods, hedge-banks. T, ex Ir, Ic, Fi, No, (but Br). [2]

Changing Forget-me-not *Myosotis discolor* BORAGINACEAE H 8–25 cm. Erect, hairy, annual herb. Lvs <4 cm, lanceolate. Fls <2 mm, tubular, yellow or white, becoming blue with age, May–Sep. Frs shining dark brown or almost black nutlets. Grassland, cultivation, on light soils. T, ex Ic, Fi. [3]

Field Gromwell *Lithospermum arvense* BORAGINACEAE H 10–50 cm. Erect, hairy, annual herb, sparsely branched. Lvs <3 cm, ovate to linear, narrower towards stem, upper sessile, lower stalked. Fls <9 mm, funnel-shaped, white, rarely purple or blue, in short, terminal clusters, May–Sep. Frs grey-brown, warty nutlets. Arable fields. T, ex Ic. [4]

Vervain *Verbena officinalis* VERBENACEAE H 30–60 cm. Erect, woody, square-stemmed, perennial herb, branched above. Lvs opposite, <75 mm, rhomboid in outline, lower divided into ovate to oblong lobes. Fls <4 mm, pale lilac, in slender, terminal spikes, becoming loose in fr, Jun–Sep. Frs red-brown nutlets with 4–5 longitudinal ribs. Waste places. T, ex Ic, FS, (but Ir, De). [5]

Bugle *Ajuga reptans* LABIATAE H 10–30 cm. Creeping, square-stemmed, perennial herb with long, overground runners. Lvs opposite, <9 cm, ovate, often tinged blue. Fls <17 mm, 2-lipped (upper lip very short, lower 3-lobed with central lobe notched), blue, pink or white, in clusters of 6 forming terminal spike, Apr–Jul. Damp woods, meadows. T, ex Ic, (but Fi). [6]

Ground-ivy *Glechoma hederacea* LABIATAE H 10–30 cm. Creeping, hairy, perennial herb with long, overground runners. Lvs opposite, <3 cm across, kidney-shaped, bluntly toothed. Fls <2 cm, 2-lipped (upper lip 2-lobed, lower 3-lobed with central lobe notched), violet with purple spots on lower lip, in axillary clusters of 2–4, Mar–May. Woods, hedge-banks, grassland. T, ex Ic. [7]

Selfheal *Prunella vulgaris* LABIATAE H 5–30 cm. Erect, hairy, square-stemmed, perennial herb with short underground stems. Lvs opposite, <5 cm, ovate, sparsely toothed, with stalks <4 cm. Fls <14 mm, 2-lipped (upper very concave), violet, pink or rarely white, in whorls of *c*6 forming dense, terminal spike, Jun–Oct. Grassland, open woods, waste places. T. [8]

Common Hemp-nettle *Galeopsis tetrahit* LABIATAE H 10–100 cm. Erect, hairy, annual herb with opposite, ascending branches, stems swollen at nodes. Lvs opposite, <10 cm, ovate, coarsely toothed. Fls <2 cm, 2-lipped (upper helmet-shaped, lower 3-lobed), pink, purple or white, with dark markings, Jul–Oct. Arable fields, woods, marshes, heaths. T. [9]

Large-flowered Hemp-nettle *Galeopsis speciosa* LABIATAE H 20–100 cm. Erect, hairy, annual herb similar to *G. tetrahit*, but more robust, with bigger, more handsome fls <35 mm, yellow with large violet blotch on lower lip. Arable fields. T, ex Ic, Lu. [10]

1
2
3
4
5
6
7
8
9
10

Red Dead-nettle *Lamium purpureum* LABIATAE H 10–40 cm. Erect, hairy, square-stemmed, annual herb, branching from base. Lvs opposite, <5 cm, ovate, often tinged red. Fls <15 mm, 2-lipped, pinkish-purple, in dense, axillary whorls, Mar–Dec. Cultivated ground, waste places. T, (but Ic). [1] White dead-nettle *L. album*, similar, but fls white. T, (but Ic). [2]

Yellow Archangel *Lamiastrum galeobdolon* LABIATAE H 20–60 cm. Erect or ascending, perennial herb. Lvs opposite, <7 cm, ovate. Fls <2 cm, 2-lipped (upper helmet-shaped, lower 3-lobed), yellow with brownish markings, Apr–Jun. Woods, hedgerows. T, ex Ic, (but Fi, No). [3]

Black Horehound *Ballota nigra* LABIATAE H 40–120 cm. Erect, hairy, square-stemmed, perennial herb. Lvs opposite, <5 cm, ovate or roundish, coarsely toothed, stalked. Fls <18 mm, 2-lipped (upper somewhat concave, lower 3-lobed, central lobe notched), purple, in many-fld, axillary whorls, May–Sep. Roadsides, waste places. T, ex Ic, Fi, No. [4]

Field Woundwort *Stachys arvensis* LABIATAE H 10–25 cm. Ascending, hairy, square-stemmed, annual herb, branching from base. Lvs opposite, <3 cm, ovate, toothed, stalked. Fls <7 mm, 2-lipped (upper somewhat concave, but not helmet-shaped, lower 3-lobed), pale purple, in axillary whorls of 2–6 in loose spike, Apr–Oct. Sandy fields. T, ex Ic, Fi, No. [5]

Hedge Woundwort *Stachys sylvatica* LABIATAE H 30–100 cm. Very hairy, erect, square-stemmed, perennial herb. Lvs opposite, <9 cm, broadly ovate, heart-shaped at base. Fls <15 mm, 2-lipped (upper somewhat concave, lower 3-lobed, central lobe notched), claret-coloured, in axillary whorls of *c*6, Jun–Sep. Woods, hedges. T, ex Ic. [6]

Betony *Betonica officinalis* LABIATAE H 15–100 cm. Erect, slightly hairy, rosette-forming, perennial herb. Lvs opposite, <7 cm, oblong, blunt, coarsely but bluntly toothed; basal lvs long-stalked, <7 cm. Fls <15 mm, 2-lipped (upper nearly flat, lower 3-lobed, central lobe notched), bright red-purple, in whorls in dense, terminal spike, Jun–Sep. Grassland, heaths, woods. T, ex Ic, (but Fi, No). [7]

Meadow Clary *Salvia pratensis* LABIATAE H 30–100 cm. Erect, somewhat hairy, square-stemmed, branched, perennial herb. Lvs opposite, <15 cm, ovate, blunt, wrinkled. Fls <25 mm, 2-lipped (upper concave, laterally flattened, forming hood), violet-blue, in whorls of *c*6 in long, sticky, terminal spike, May–Jul. Grassland, roadsides. Br, Fr, Lu, Be, Ne, Ge, Cz, Po, (Sw). [8]

Wild Basil *Clinopodium vulgare* LABIATAE H 30–80 cm. Erect, hairy, square-stemmed, perennial herb. Lvs opposite, <5 cm, ovate, scarcely toothed, short-stalked. Fls <2 cm, unequally 4-lobed, rose-purple, in whorls, Jul–Sep. Open woods, hedges, dry grassland. T, ex Ic. [9]

Marjoram *Origanum vulgare* LABIATAE H 30–80 cm. Erect, hairy, scented, perennial herb, branching above. Lvs opposite, <45 mm, ovate. Fls <8 mm, almost equally 4-lobed, rose-purple, Jun–Sep. Open woods, hedges, dry grassland. T, ex Ic. [10]

Large Thyme *Thymus pulegioides* LABIATAE H 10–25 cm. Trailing, strongly scented, perennial herb, somewhat woody at base, hairy on angles of square stems. Lvs <18 mm, ovate. Fls <6 mm, 2-lipped (upper broad and notched, lower 3-lobed), pink-purple, Jul–Sep. Dry grassland. T, ex Ir, Ic, (but Fi). [11]

Corn Mint *Mentha arvensis* LABIATAE H 10–60 cm. Erect or ascending, hairy, perennial herb. Lvs <5 cm, elliptical, blunt, shallowly toothed. Fls <8 mm long, with 4 nearly equal lobes, lilac or white, May–Oct. Arable, open woods, damp grassland. Not pungently scented. T, ex Ic. [12]

Bittersweet *Solanum dulcamara* SOLANACEAE H <2 m. Scrambling, woody perennial. Lvs <8 cm, ovate, often with 1–4 deep lobes on stalked segments at base. Fls purple with yellow centre, <15 mm across, Jun–Sep. Frs red berries. Hedgerows, wet woods. T, ex Ic. [1]

Black Nightshade *Solanum nigrum* SOLANACEAE H 10–70 cm. Prostrate or erect, annual herb. Lvs <7 cm, ovate to lanceolate, short-stalked. Fls <15 mm, white with yellow anthers, Jun–Sep. Frs green or black berries, <8 mm, poisonous. Waste places. T, ex Ic. [2]

Great Mullein *Verbascum thapsus* SCROPHULARIACEAE H 30–200 cm. Erect, rosette-forming, biennial herb, covered with soft, whitish wool. Basal lvs <45 cm, lanceolate, narrowing to winged stalk; stem lvs smaller. Fls <3 cm, yellow with white-haired anthers, of 5 petals, in dense, terminal spike, Jun–Sep. Frs ovoid capsules, <1 cm. Dry, waste places. T, ex Ic. [3]

Common Toadflax *Linaria vulgaris* SCROPHULARIACEAE H 30–80 cm. Erect, yellow-green, hairless, perennial herb, branched above. Lvs <8 cm, narrow. Fls <25 mm long, yellow, 2-lipped, with long spur, lower lip 3-lobed, with bulging, orange portion in centre; in dense, many-fld, elongate spike, Jun–Oct. Frs ovoid capsules, <11 mm. Hedge-banks, waste places. T, ex Ic. [4]

Small Toadflax *Chaenorhinum minus* SCROPHULARIACEAE H 8–25 cm. Erect, glandular-hairy, annual herb, with ascending branches. Lvs <25 mm, linear-lanceolate, narrowed to short stalk. Fls <8 mm, purple outside, paler within, often tinged yellow, 2-lipped, with short spur, May–Oct. Frs ovoid capsules, <6 mm. Dry places. T, ex Ic. [5]

Common Figwort *Scrophularia nodosa* SCROPHULARIACEAE H 40–80 cm. Erect, hairless, square-stemmed, perennial herb. Lvs opposite, <13 cm, ovate, acute, coarsely toothed. Fls <1 cm, green and red-brown, Jun–Sep. Frs ovoid capsules, <1 cm. Woods, hedge-banks. T, ex Ic. [6]

Thyme-leaved Speedwell *Veronica serpyllifolia* SCROPHULARIACEAE H 10–30 cm. Creeping, perennial herb, often forming dense mats, with ascending, flowering stems. Lvs opposite, <2 cm, oval. Fls <1 cm, of 4 white or pale blue petals with dark lines, in terminal spikes of <30, Mar–Oct. Frs hairy capsules, <5 mm. Grassland, heaths. T. [7]

Germander Speedwell *Veronica chamaedrys* SCROPHULARIACEAE H 20–40 cm. Creeping, hairy, perennial herb. Lvs opposite, <25 mm, triangular-ovate. Fls <1 cm, of 4 bright blue petals, with white centre, Mar–Jul. Frs hairy, heart-shaped capsules, <5 mm. Grassland, woods, hedges. T, (but Ic). [8]

Foxglove *Digitalis purpurea* SCROPHULARIACEAE H 50–150 cm. Erect, hairy, rosette-forming, biennial or, rarely, perennial herb. Lvs <30 cm, ovate to lanceolate, with rounded teeth. Fls <5 cm long, bell-shaped, pink-purple, in spikes of 20–80, Jun–Sep. Frs ovoid capsules, <15 mm. Woods, hedges. T, ex Ic, Fi, (but Ne, De, Po). [9]

Greater Yellow Rattle *Rhinanthus serotinus* SCROPHULARIACEAE H 20–60 cm. Erect, annual herb with black-spotted stem. Lvs opposite, <7 cm, lanceolate, toothed, stalkless. Fls <2 cm long, 2-lipped, yellow, with 2 violet teeth on upper lip, Jun–Sep. Frs capsules, <7 mm, with seeds rattling inside. Meadows, arable fields. T, ex Ir, Ic. [10]

Common Cow-wheat *Melampyrum pratense* SCROPHULARIACEAE H 8–60 cm. Very variable, erect, annual herb. Lvs opposite, <10 cm, lanceolate. Fls <17 mm long, 2-lipped, deep yellow to white, in axils of lfy bracts, Jun–Oct. Frs flattened, 4-seeded capsules. Woods, semi-parasitic on grasses. T, ex Ic. [11]

1
2
3
4
5
6
7
8
9
10
11

Ribwort Plantain *Plantago lanceolata* PLANTAGINACEAE H <40 cm. Usually hairless, perennial herb. Lvs <15 cm, in rosette, lanceolate, ribbed. Fls <4 mm, brownish, of 4 petals, in spike on long stalk, Apr–Aug. Ubiquitous in grassland, except far north. Size very variable: in dry grassland, some plants have lvs only 1–2 cm long. T. [1]

Elder *Sambucus nigra* CAPRIFOLIACEAE H <10 m. Deciduous shrub or small tree with arching, hollow branches and deeply furrowed, corky bark. Lvs opposite, divided into <7 ovate, toothed segments <12 cm. Fls <5 mm, of 5 creamy-white petals, in flat-topped umbels <20 cm across, Jun–Aug. Frs black, spherical berries, <8 mm. Woods, scrub, hedgerows. Fls and frs used for wine, but lvs and unripe frs poisonous. T, ex Ic, (but FS). [2]

Guelder-rose *Viburnum opulus* CAPRIFOLIACEAE H <4 m. Small, deciduous shrub with smooth, greyish twigs. Lvs opposite, <8 cm, with irregularly toothed lobes, reddening in autumn. Fls white, in flat-topped clusters <10 cm across, fertile inner fls <6 mm across, sterile outer <2 cm, Jun–Jul. Frs red, spherical berries, <8 mm. Woods, scrub, hedgerows. Bark, lvs and frs poisonous to man, but birds eat berries with impunity. T, ex Ic. [3]

Honeysuckle *Lonicera periclymenum* CAPRIFOLIACEAE H <6 m. Rampant, twining, deciduous shrub with hollow branches. Lvs opposite, <7 cm, elliptical, dark green above, pale beneath, short-stalked. Fls <5 cm long, funnel-shaped, 2-lipped, creamy-white inside, purplish outside, in terminal clusters, Jun–Sep. Frs spherical, red berries, <5 mm. Woods, hedges. Strongly scented at night, attracts moths. T, ex Ic, Fi, (but Cz). [4]

Teasel *Dipsacus fullonum* ssp *sylvestris* DIPSACACEAE H 50–200 cm. Erect, biennial herb with prickly, angled stems. Stem lvs opposite, <30 cm, narrowly lanceolate, joined at base into water-collecting cup. Fls <11 mm long, rose-purple, of 4 unequal petals, in blunt, conical, spiny heads <8 cm, Jul–Aug. Open woods, banks, waste places, esp on clay. T, ex Ic, FS, (but De). [5]

Field Scabious *Knautia arvensis* DIPSACACEAE H 25–100 cm. Variable, erect, branched, rosette-forming, perennial herb; stem rough below, with down-turned bristles. Stem lvs opposite, <20 cm, divided into <12 oblong, lateral lobes and elliptical, terminal lobe. Fls <1 cm long, bluish-lilac, in heads <4 cm across, outer fls larger, Jul–Sep. Dry grassland, banks, open woods. T, (but Ic). [6]

Clustered Bellflower *Campanula glomerata* CAMPANULACEAE H 15–80 cm. Erect, hairy, perennial herb. Lvs <4 cm, ovate-lanceolate, obtusely toothed, lower long-stalked. Fls <25 mm, bell-shaped, blue, in dense, terminal clusters, May–Sep. Frs capsules, <3 mm. Scrub, dry grassland. T, ex Ir, Ic, (but No). [7]

Peach-leaved Bellflower *Campanula persicifolia* CAMPANULACEAE H 40–100 cm. Slender, erect perennial herb. Lvs <10 cm, linear-lanceolate, bluntly and sparsely toothed, lower stalked. Fls <4 cm, open funnel-shaped, blue, in spikes of 2–8, May–Aug. Frs 10-veined capsules, <5 mm. Meadows, open woods. T, ex Ir, Ic, (but Br). [8] Spreading bellflower *C. patula*, similar in habitat and distribution, has smaller fls, <25 mm, in much-branched, loose, spreading clusters.

Creeping Bellflower *Campanula rapunculoides* CAMPANULACEAE H 30–100 cm. Erect, almost hairless, perennial herb, spreading by an extensive, underground root system. Lower lvs <8 cm, ovate or heart-shaped, long-stalked; upper lvs narrower, stalkless. Fls <30 mm, funnel-shaped, blue-purple, nodding, Jul–Sep. Frs round, nodding capsules, <7 mm. Open woods, grassy places. T, ex Ic, (but Br, Ir). [9]

3
4
2
5
6
8
1
7
9

Yarrow *Achillea millefolium* COMPOSITAE H 8–45 cm. Creeping, aromatic, perennial herb with erect, furrowed, woolly stems. Lvs 5–15 cm, dark green, feathery. Fl-heads white or pink, 4–6 mm across, numerous in dense, terminal clusters, Jun–Aug. Grassland. T. [1]

Ox-eye Daisy *Chrysanthemum leucanthemum* COMPOSITAE H 20–70 cm. Erect, rosette-forming, perennial herb. Lower lvs <10 cm, rounded, broadest at tip, toothed, long-stalked; upper smaller, stalkless. Fl-heads <5 cm, white with yellow centre, solitary, long-stalked, May–Aug. Grassland. T, (but Ic). [2]

Mugwort *Artemisia vulgaris* COMPOSITAE H 60–120 cm. Erect, aromatic, perennial herb, tufted at base, with reddish stems. Lvs <8 cm, much-divided into lanceolate segments, dark green and hairless above, whitish and hairy beneath. Fl-heads ovoid, <2·5 mm, with red-brown florets, in dense clusters, Jul–Sep. Waste places, roadsides. T, ex Ic. [3]

Colt's-foot *Tussilago farfara* COMPOSITAE H <15 cm. Creeping perennial with long, white, scaly, prostrate stems. Lvs <20 cm across, roundish-polygonal (like colt's foot), appearing after fls. Fl-heads yellow, 15–35 mm across, solitary, terminal, on erect, scaly, woolly shoots 5–15 cm, Mar–Apr. Waste places, esp clay soils. T. [4]

Common Ragwort *Senecio jacobaea* COMPOSITAE H 30–150 cm. Erect, biennial or perennial herb. Lvs dark green, simply or twice pinnate, variously toothed and lobed. Fl-heads <25 mm across, yellow, in dense, flat-topped clusters, Jun–Oct. Waste places, grassland, esp overgrazed pastures; rare in far north. T, ex Ic, Fi. [5]

Lesser Burdock *Arctium minus* COMPOSITAE H 60–130 cm. Much-branched, biennial herb with stout tap-root. Lvs <50 cm, broadly ovate, with hollow stalk. Fl-heads ovoid, <25 mm, with purple florets, overtopping long, hooked bracts, Jul–Sep. Frs brownish, 1-seeded, <7 mm. Woods, hedges, waste places. T, ex Ic. [6]

Musk Thistle *Carduus nutans* COMPOSITAE H 20–100 cm. Much-branched, biennial herb with spiny, winged stem, naked only beneath fl-heads. Basal lvs elliptical, <30 cm; upper divided into 2–5 lobed, spine-tipped segments. Fl-heads red-purple, <5 cm, drooping, in loose clusters of 2–4, May–Aug. Frs fawn, 1-seeded, <4 mm. Grassland, roadsides, waste places. Br, Fr, Lu, Be, Ne, Ge, Cz, (De, Sw). [7] Creeping thistle *Cirsium arvense* also has stem naked beneath fl-heads, but heads small, <25 mm, pale purple or white, spreads by far-creeping underground stems; spear thistle *C. vulgare* has stems completely winged and large, red-purple fl-heads, <5 cm. Both T.

Cat's-ear *Hypochoeris radicata* COMPOSITAE H 20–60 cm. Erect, branched, rosette-forming, perennial herb. Lvs <25 cm, oblong, broadest near tip, lobed, sparsely hairy. Fl-heads <3 cm, yellow, solitary, on loose, lfless branches, Jun–Sep. Frs orange-beaked, 1-seeded, <7 mm. Grassland, waste places. T, ex Ic, Fi. [8]

Dandelion *Taraxacum officinale* COMPOSITAE H <40 cm. Erect, rosette-forming, perennial herb with deep tap-root, lvs and stems producing milky sap when cut; includes aggregate of similar microspp, difficult to separate. Lvs <40 cm, coarse, variously divided. Fl-heads yellow, <6 cm across, solitary, terminal, Apr–Oct. Frs familiar 'clocks'. Roadsides, wasteland, woods, meadows. T. [9]

Prickly Sow-thistle *Sonchus asper* COMPOSITAE H 20–150 cm. Erect, hairless, annual or biennial herb, stems producing copious milky sap when cut. Lvs glossy-green, pinnate, those on stem with rounded, ear-like flaps at base. Fl-heads golden-yellow, 20–25 mm across, in irregular clusters, Jun–Aug. Fluffy frs dispersed by wind. Waste places, dry banks. T, ex Ic. [10]

7
3
10
6
5
1
8
2
9
4

Meadow Saffron *Colchicum autumnale* LILIACEAE H 10–30 cm. Hairless, perennial herb, producing large corms, <5 cm across. Lvs in spring, <30 cm, oblong-lanceolate, shining green. Fls <20 cm long, of 6 pale purple perianth segments, appearing after lvs die down, Aug–Sep. Frs ovoid capsules, <5 cm, appearing from below ground with lvs in spring. Poisonous to grazing animals and, for this reason, threatened with extinction in many areas. Damp meadows, woods. T, ex Ic, Fi, (but De, Sw, No). [1]

Wild Onion *Allium vineale* LILIACEAE H 30–80 cm. Erect, bulbous, perennial herb. Lvs <60 cm, cylindrical, grooved, hollow. Fls <5 mm long, bell-shaped, of 6 reddish perianth segments, in loose, terminal umbel of bulbils and fls, or more compact umbel of bulbils with few or no fls, surrounded by greenish-white spathe, on long stalk <80 cm, Jun–Jul. Meadows, roadsides. Once serious weed of pastures, tainting milk. T, ex Ic. [2]

Star-of-Bethlehem *Ornithogalum umbellatum* LILIACEAE H 10–30 cm. Hairless, bulbous, perennial herb. Lvs <30 cm, linear, with white-striped midrib. Fls <2 cm, of 6 white perianth segments with green stripe on back, in umbel of 5–30, Apr–Jun. Cultivated ground, open grassland; grown in gardens. T, ex Ir, Ic, (but FS). [3]

May Lily *Maianthemum bifolium* LILIACEAE H 8–15 cm. Erect, perennial herb, with stiff, appressed hairs above; long, creeping, underground stems. Lvs usually 2, <6 cm, ovate, acute, heart-shaped at base. Fls tiny, <4 mm, of 4 perianth segments, in dense, terminal clusters of 8–15, May–Jul. Frs red berries, <6 mm. Woods. T, ex Ir, Ic. [4]

Solomon's-seal *Polygonatum multiflorum* LILIACEAE H 30–80 cm. Arching, perennial herb with long, creeping, underground stems. Lvs <12 cm, elliptical, stalkless, almost in 2 rows. Fls <15 mm, contracted in middle, of 3 greenish-white perianth segments, in stalked clusters of 2–5, odourless, Apr–Jun. Frs blue-black berries, <8 mm. Woods. T, ex Ir, Ic. [5]

Bluebell *Endymion non-scriptus* LILIACEAE H 20–50 cm. Erect, bulbous, perennial herb. Lvs <45 cm, linear, shiny green. Fls <2 cm long, bell-shaped, blue (rarely pink or white), in terminal clusters of 4–20, opening from bottom upwards, sweet-scented, Apr–May. Frs ovoid capsules, <15 mm. Woods, thickets, hedges. Br, Ir, Fr, Be, Ne, (Ge). [6]

Lily-of-the-valley *Convallaria majalis* LILIACEAE H 10–25 cm. Hairless, creeping, perennial herb. Lvs in pairs, <20 cm, broadly elliptical, glossy, acute. Fls <8 mm, white, bell-shaped, nodding on 1-sided, terminal spikes, fragrant, May–Jun. Frs red berries. Dry woods. T, ex Ic. [7]

Herb-Paris *Paris quadrifolia* TRILLIACEAE H 15–40 cm. Erect, perennial herb with long, creeping underground stems. Lvs in single whorl of 4–6, <12 cm, ovate. Fls <35 mm, of 4–6 greenish outer perianth segments, and 4–6 yellowish inner segments, May–Jul. Frs spherical black berries, <15 mm, poisonous. Damp woods on calcareous soils. T, ex Ir. [8]

Snowdrop *Galanthus nivalis* AMARYLLIDACEAE H 15–25 cm. Bulbous, hairless, perennial herb. Lvs in pairs, <25 cm long, narrow, grey-green. Fls 20–25 mm long, bell-shaped, white, with 3 small inner petals tipped green, nodding, solitary, on long stalks, Jan–Mar. Frs ovoid capsules. Woods, plantations, often naturalized. T, ex Ic. [9]

Black Bryony *Tamus communis* DIOSCOREACEAE H <4 m. Climbing, perennial herb with black, underground tuber. Lvs <10 cm, heart-shaped, dark shining green. Fls <5 mm, of 6 yellow-green perianth segments, in spikes, ♂ and ♀ on separate plants, May–Jul. Frs pale red berries, <12 mm, poisonous. Wood margins, hedges. Br, Fr, Be, Ge. [10]

Hairy Wood-rush *Luzula pilosa* JUNCACEAE H 20–40 cm. Lvs <20 cm, sparsely hairy. Fls <6 mm, of 6 dark chestnut-brown segments, in loose clusters, Apr–Jun. Woods. T, ex Ic. [1]

Great Wood-rush *Luzula sylvestris* JUNCACEAE H 40–80 cm. Lvs <30 cm, glossy, sparsely hairy. Fls <6 mm, of 6 chestnut-brown segments, Apr–Jul. Woods. T, ex Ic, Fi. [2]

Field Wood-rush *Luzula campestris* JUNCACEAE H 10–30 cm. Lvs <10 cm, with sparse, long, white hairs. Fls <6 mm, of 6 chestnut-brown segments with transparent margins, Mar–Jun. Grassland. T, ex Ic. [3]

Soft Brome *Bromus mollis* GRAMINEAE H 5–80 cm. Lvs flat, <7 mm wide, softly hairy. Fls awned, in rotund spikelets of 6–12, <2 cm, in dense panicles, May–Aug. Meadows, wasteland. T. [4]

Upright Brome *Bromus erectus* GRAMINEAE H 60–100 cm. Lvs flat, <1 cm wide, hairy. Fls awned, in spikelets of 4–14, <4 cm, in loose, erect panicles, May–Jul. Calcareous grassland. T, ex Ic, (but Ir). [5]

Hairy Brome *Bromus ramosus* GRAMINEAE H 100–140 cm. Down-turned hairs at base of stem. Lvs flat, <1 cm wide. Fls awned, in spikelets of 4–11, <4 cm, in nodding panicles, Jun–Aug. Woods, hedgerows. T, ex Ic, Fi. [6]

Tor-grass *Brachypodium pinnatum* GRAMINEAE H 30–120 cm. Lvs rolled at base, <5 mm wide, yellow-green. Fls short-awned, in spikelets of 8–22, <4 cm, erect, stalkless, in spike-like panicles, Jun–Aug. Calcareous grassland. T, ex Ic, (but Ir). [7]

False Brome *Brachypodium sylvaticum* GRAMINEAE H 30–90 cm. Lvs flat, <13 mm wide, hairy, yellow-green, drooping. Fls long-awned, <12 mm, in spikelets of 8–16, <4 cm, stalkless, in nodding, spike-like panicles. T, ex Ic, Fi. [8]

Wall Barley *Hordeum murinum* GRAMINEAE H 30–60 cm. Lvs flat, <8 mm wide, hairy. Fls long-awned, <3 cm, in dense spikes, <12 cm long, May–Aug. Waste places. T, ex Ic. [9]

Wood Melick *Melica uniflora* GRAMINEAE H 30–70 cm. Lvs flat, <7 mm wide, hairy. Fls awnless, in 1-fld spikelets, <7 mm, in erect or nodding panicles, May–Jul. Woods, shady banks. T, ex Ic, Fi. [10]

Cock's-foot *Dactylis glomerata* GRAMINEAE H 20–100 cm. Lvs folded, later flat, <14 mm wide. Fls short-awned, in stalkless spikelets of 2–5, <9 mm, on 1-sided, branched panicle, May–Aug. T. [11]

Tall Fescue *Festuca arundinacea* GRAMINEAE H 60–200 cm. Lvs flat, <1 cm wide. Fls awnless, in long-stalked spikelets of 3–10, <18 mm, in erect or nodding panicles, Jun–Aug. T, ex Ic. [12]

Giant Fescue *Festuca gigantea* GRAMINEAE H 50–150 cm. Lvs flat, <18 mm wide. Fls long-awned, <18 mm, in spikelets of 3–10, <2 cm, in loose, nodding panicles, Jun–Aug. Woods, shady banks. T, ex Ic. [13]

Sheep's-fescue *Festuca ovina* GRAMINEAE H 10–50 cm. Lvs rolled, <1 mm wide. Fls short-awned, <1·5 mm, in spikelets of 3–9, <1 cm, in narrow, oblong panicles, May–Jul. Dry grassland. T. [14]

Wild-oat *Avena fatua* GRAMINEAE H 30–150 cm. Lvs flat, <15 mm wide; basal lvs usually hairy. Fls long-awned, <4 cm, in spikelets of 2–3, <25 mm, drooping in loose panicles, Jun–Aug. Arable fields, waste places. T, ex Ic. [15]

False Oat-grass *Arrhenatherum elatius* GRAMINEAE H 50–150 cm. Fls awned in spikelets of 2, in loose panicles, <30 cm, Jun–Jul. Meadows, roadsides. T.

Downy Oat-grass *Helictotrichon pubescens* GRAMINEAE H 30–100 cm. Lvs folded, later flat, <6 mm wide, hairy. Fls with long, twisted awns, <2 cm, in spikelets of 2–3, <13 mm, in erect or nodding panicles, May–Jul. T, ex Ic. [16]

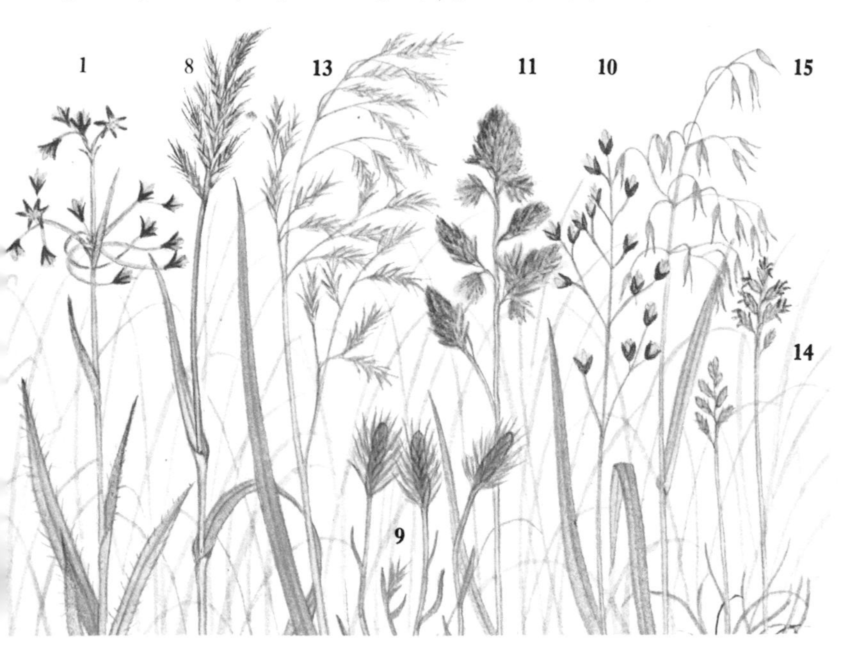

Common Bent *Agrostis tenuis* GRAMINEAE H 20–50 cm. Lvs flat, <3 mm wide. Fls tiny, awnless, in 1-fld spikelets, <3 mm, in very loose, erect, pyramidal panicles, Jun–Aug. Heaths, moors, acid grassland. T. [1]

Meadow Foxtail *Alopecurus pratensis* GRAMINEAE H 30–120 cm. Lvs flat, <1 cm wide. Fls long-awned, <5 mm, in 1-fld spikelets, <6 mm, in dense, narrow, cylindrical spike <13 cm long, Apr–Jun. Damp grassland. T, (but Ic). [2]

Wood Millet *Milium effusum* GRAMINEAE H 50–150 cm. Lvs flat, <1 cm wide. Fls awnless, in 1-fld spikelets, <4 mm, in loose, nodding, wavy-branched, pyramidal panicles with main branches turning down, May–Jul. Woods. T. [3]

Sweet Vernal-grass *Anthoxanthum odoratum* GRAMINEAE H 20–50 cm. Lvs flat, <5 mm wide, sparsely hairy. Fls awned, <9 mm, in spikelets of 3, <1 cm, in moderately dense, spike-like panicles <6 cm long, Apr–Jul. Meadows, heaths, moors. Gives fragrance to new-mown hay. T. [4]

Spring-sedge *Carex caryophyllea* CYPERACEAE H 5–15 cm. Creeping, with slender, 3-angled stems. Lvs flat, <2 mm wide. Fls in 3–4 stalkless, separate ♂ and ♀ spikes, <15 mm; single ♂ above, with conspicuous yellow anthers clustered at top of erect stem, Apr–Jul. Dry grassland. T, ex Ic. [5]

Wood-sedge *Carex sylvatica* CYPERACEAE H 15–60 cm. Stems smooth, 3-angled. Lvs slightly keeled, <6 mm wide. Fls in 4–6 stalked, separate ♂ and ♀ spikes <5 cm, single ♂ above, ♀s below, well-spaced and nodding, May–Jul. Woods. T, ex Ic, Fi. [6]

Remote Sedge *Carex remota* CYPERACEAE H 30–60 cm. Stems bluntly 3-angled. Lvs channelled, <2 mm wide, bright green. Fls in 4–7 stalkless spikes, <1 cm, widely spaced in axils of long, lf-like bracts, May–Jul. Wet woods. T, ex Ic, Fi. [7]

Lords-and-Ladies *Arum maculatum* ARACEAE H 30–50 cm. Erect, hairless, perennial herb with underground tuber. Lvs <20 cm, triangular, often black-spotted, long-stalked. Erect, pale-green spathe, opening to reveal dull purple or yellow spadix (fleshy fl-stem) with separate ♂ and ♀ fls at base, Apr–May. Frs scarlet berries, <5 mm, poisonous. Woods, hedge-banks. T, ex Ic, Fi, No. [8]

Early Purple Orchid *Orchis mascula* ORCHIDACEAE H 15–60 cm. Erect, hairless, perennial herb with ovoid tubers. Lvs <2 cm, oblong, blunt, usually black-spotted. Fls <12 mm, dark crimson-purple, with stout, horizontal spur, 2-lipped, lower lip 3-lobed and strongly spotted, in cylindrical spike, <15 cm, Apr–Jul. Woods, copses, meadows. T, ex Ic, Fi. [9]

Greater Butterfly Orchid *Platanthera chlorantha* ORCHIDACEAE H 20–60 cm. Erect, hairless, perennial herb. Lower lvs <15 cm, elliptical; upper lvs smaller, bract-like. Fls <23 mm across, whitish, with long, slender spur <28 mm, in loose pyramidal spike, <20 cm, strongly night-scented, May–Jul. Woods, grassland. T, ex Ic. [10]

Broad-leaved Helleborine *Epipactis helleborine* ORCHIDACEAE H 25–80 cm. Perennial herb with 1–3 erect stems, violet-tinged below. Lvs <17 cm, prominently 5-veined. Fls <1 cm across, greenish or dull purple, with 'helmet' above, in dense spike of 15–50, <30 cm, Jul–Sep. Woods. T, ex Ic. [11]

Bird's-nest Orchid *Neottia nidus-avis* ORCHIDACEAE H 20–45 cm. Erect, saprophytic, glandular-hairy, perennial herb. No green lvs; stem covered in brownish scales. Fls <12 mm across, brownish, with lower lip divided into 2 blunt lobes, in dense spikes of 40–60, May–Jul. Woods. T, ex Ic. [12]

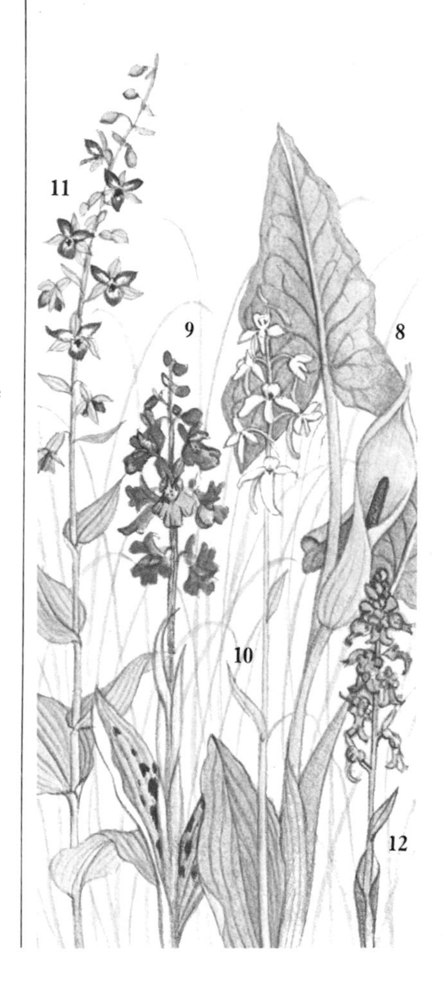

Slugs and snails mostly herbivorous; some feed on carrion, even attack invertebrates. Hermaphrodite, but mate to exchange sperm. Courtship sometimes elaborate, esp in slugs; may mate for several hours. Eggs laid in batches on ground; hatch in 3–5 weeks; young miniature adults.

Large Black Slug *Arion ater* var *ater*
ARIONIDAE EL *c*14 cm. Very large; no keel on back, prominent mantle, respiratory orifice; colour variable; juv straw-yellow with faint grey bands. Mucus sticky, colourless to orange. Common in woods, grassland, waste places generally. Breeds most of year. Omnivorous. T. [1] Large red slug *Arion ater* var *rufus*, very similar, but generally redder. [2]

Netted Slug *Agriolimax reticulatus*
LIMACIDAE EL <35 mm. Small; short keel on back; whitish or brown-yellow to blue-black, irregularly mottled with grey or dark brown. Mucus chalky white. Common in grassland, waste places, wooded areas, with thick ground vegetation. Breeds throughout year. T. [3]

Rounded Snail *Discus rotundatus*
ENDODONTIDAE SH 2·5–3 mm, SB 6–7 mm, W 6–7. Shell dextral, spire slightly raised, whorls cylindrical, with slight peripheral keel; numerous, strong, transverse ridges; umbilicus large. Shell yellow-brown with transverse reddish blotches. Common in woodlands, also grassland, even on acid soils. Breeds all year. T, ex Ic, nFS. [4]

Plaited Door Snail *Cochlodina laminata*
CLAUSILIIDAE SH 15–18 mm, SB 4 mm, W 11–12. Shell sinistral, club-shaped with tapering spire; glossy, with fine, transverse striations; aperture oval, with pale peristome; 2 prominent tooth-like folds inside shell, 3–4 folds deeper within. Shell yellowish-brown, sometimes reddish. Woodland, under logs, occasionally up trees. T, ex nBr, Ic, nFS. [5]

Door Snail *Clausilia bidentata*
CLAUSILIIDAE SH 11–15 mm, SB 3–3·5 mm,

W 11–13. Shell sinistral, slender, slightly glossy with numerous wavy striations; aperture irregularly oval, with pale backward-curving peristome, 2 prominent tooth-like folds inside shell. Shell grey-brown with interrupted, transverse, white streaks; animal dark grey. Common under fallen wood, in litter, in woodlands, hedges. T, ex Ic, Po, nFS. [6]

Grove Snail *Cepaea nemoralis* HELICIDAE SH 15–20 mm, SB 17–26 mm, W 5½. Shell dextral, glossy; aperture large, rounded, with internal thickening; umbilicus closed. Shell yellow, pink or brown, with dark peristome and 0–5 dark spiral bands; animal yellowish-grey. Frequent in woods, grasslands. Often eaten by birds, small mammals. T, ex Ic, nFS. [7]

Glossy Glass Snail *Oxychilus helveticus* ZONITIDAE SH 4·5–5 mm, SB 9–10 mm, W 5. Shell similar to smooth glass snail, but spire more pointed, umbilicus small. Shell rich horn-brown, translucent, paler towards umbilicus; animal bluish-grey, mantle appearing as distinct dark band through shell. Frequent in woods, hedges, moist places. Br, Ir, Fr, Be. [8]

Clear Glass Snail *Aegopinella pura* ZONITIDAE SH 2–2·5 mm, SB 4–4·5 mm, W 3½–4. Shell dextral, convex above and below; semi-transparent, with fine radial and spiral striations giving silky appearance. Shell pale; animal yellowish-grey, mantle white with dark spotting. Frequent among dead leaves, moss, in damp woods, hedges. T, ex Ic, nFS. [9]

Smooth Glass Snail *Aegopinella nitidula* ZONITIDAE SH 5–6 mm, SB 9–10 mm, W 5. Shell dextral, slightly convex above and below; dull, waxy surface, with irregular transverse and fine spiral striations; umbilicus wide and deep. Shell brown, paler towards umbilicus; animal dark grey above, paler below. Common in woodland, waste places generally, less frequent grassland. Breeds late spring–summer. Mainly herbivorous, but eats some animal matter. T, ex Ic, nFS. [10]

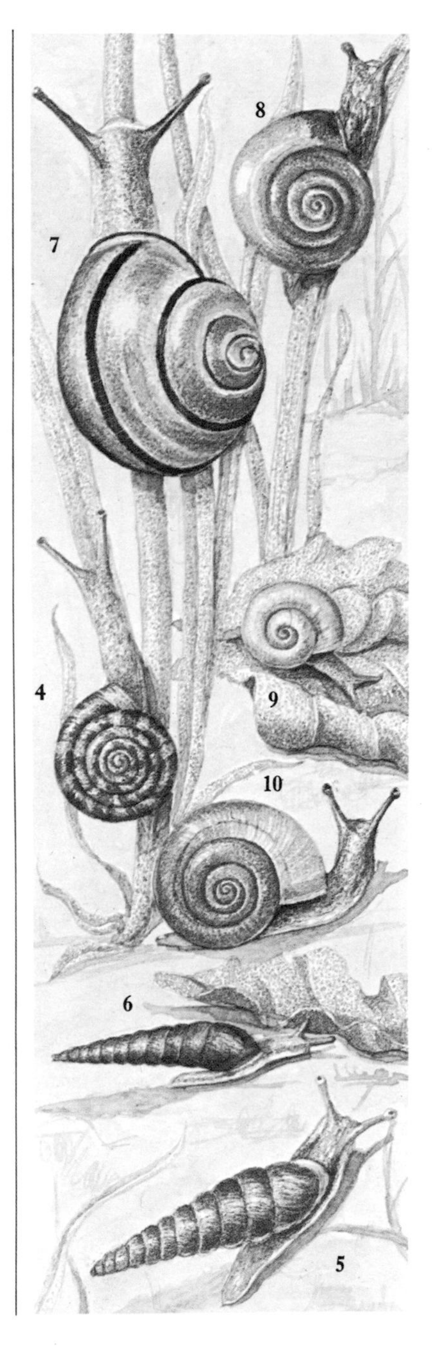

Crystal Snail *Vitrea crystallina*
ZONITIDAE SH 1·5 mm, SB 3–3·5 mm, W 4½–5. Shell dextral, more convex below; thin, shiny, with minute, irregular, transverse striations; aperture rounded, with sharp peristome and internal rib in adult; umbilicus narrow. Shell colourless; animal grey-white. Common in damp woods, grassland. T, ex Fi, nSC. [1]

Hollowed Glass Snail *Zonitoides excavatus* ZONITIDAE SH 3 mm, SB 6–8 mm, W 5½–6. Shell dextral, convex above and below, semi-transparent, shiny, with strong, irregular, transverse striations; aperture small, rounded; umbilicus large. Shell horn-brown; animal grey, mantle spotted with white. In woods on acid soils. Br, Ir, Be, Ne, De, Ge. [2]

Wrinkled Snail *Candidula intersecta*
HELICIDAE SH 6–8 mm, SB 9–12 mm, W 4½–5. Shell dextral, with strong, transverse striations; aperture rounded, peristome thin with strong internal thickening; umbilicus large. Shell white or cream with broken, brownish-purple, spiral bands. Grassland on calcareous soils. WE. [3]

Roman or Edible Snail *Helix pomatia*
HELICIDAE SH 40–45 mm, SB 40–45 mm, W 5. Shell dextral with coarse, irregular, transverse striations; aperture rounded, with thickened peristome; no umbilicus. Shell grey-brown, with spiral bands. Open woods, grassland, on calcareous soil. Fr, Be, Ge, Cz, Po, (sBr, Ne, De, Sw). [4]

Striped Snail *Cernuella virgata*
HELICIDAE SH 12 mm, SB 17 mm, W 5–6. Shell dextral, shiny; aperture large, rounded, peristome sharp with internal rib; umbilicus deep. Shell white or cream, with continuous, brown, spiral band above periphery, thinner, broken bands below. Frequent in calcareous grassland, sand-dunes. WE. [5]

Kentish Snail *Monacha cantiana*
HELICIDAE SH 12 mm, SB 19 mm, W 6–6½. Shell dextral, transparent and glossy,

with fine transverse striations; aperture round, with sharp, slightly expanded peristome and internal thickening; umbilicus narrow. Shell pale with pink or brown tinge. Common in calcareous grassland. Breeds autumn. WE. [6]

Hairy Snail *Trichia hispida* HELICIDAE SH 4·5–5 mm, SB 6·5–8 mm, W 6–7. Shell dextral, shape variable, irregular transverse striations, fine recurved hairs which may be rubbed off; aperture oval with backward-curving peristome and pale internal rib; umbilicus medium to large. Shell brownish-grey. Abundant in moist woodlands, waste places. Breeds spring–autumn. T, ex Ic, nFS. [7]

Herald Snail *Carychium tridentatum* ELLOBIIDAE SH 2–2·5 mm, SB 1–1·5 mm, W 4½–5. Shell dextral, thin-walled, with fine, transverse striations; aperture with peristome folded back on itself and 3 internal teeth. Shell and animal white. Common in litter in open woodland, grassland. T, ex Ic, nFS. [8]

Slippery Snail *Cochlicopa lubrica* COCHLICOPIDAE SH 6–7 mm, SB 2·5–3 mm, W 6. Shell dextral, smooth, shiny, transparent; aperture with blunt peristome and internal thickening; no umbilicus. Shell rich brown; animal grey. Common in moss, dead leaves, in woodland, grassland, damp areas. T. [9]

Chrysalis Snail *Lauria cylindracea* VERTIGINIDAE SH 3·5–4 mm, SB 2 mm, W 7. Shell dextral, glossy with very fine striations; aperture with single tooth and thickened, white lip. Shell brown. In woodland, grassland, on stone walls, rocks, dry areas. T, ex Ic, Fi, nSC. [10]

Beautiful Grass Snail *Vallonia pulchella* VALLONIIDAE SH 1·3 mm, SB 2·5 mm, W 3½–4. Shell dextral, transparent, with fine radial striations; aperture flared, with thickened peristome; umbilicus moderately large. Shell and animal yellowish-white. Moist fields, meadows, at bases of grass stems. T, ex Ic, nFS. [11]

Earthworms sometimes numerous (<250 per square metre of good pasture land). Burrow deep. Soil-feeders, digesting organic matter, or browsers on surface vegetation and litter. Hermaphrodite, mate to exchange sperm; then separate. Eggs and sperm deposited into mucous cocoon secreted by clitellum (saddle-like band round body). Fertilization within cocoon; small worms emerge. Position of clitellum along body is characteristic of sp; 'on segments 28, 29–37' means it begins at the 28th or 29th segment, counted from the head end, and terminates at the 37th segment.

Green Worm *Allolobophora chlorotica*
LUMBRICIDAE BL 30–70 mm. Body cylindrical; green or occasionally yellow, grey or pink; clitellum pink to orange, on segments 28, 29–37. Shallow burrows esp in arable land. T, ex Ic. [1]

Long Worm *Allolobophora longa*
LUMBRICIDAE BL 90–180 mm. Body cylindrical, tail flattened; grey or brown, with anterior and upper portions of body light or dark brown; clitellum reddish, on segments 27, 28–35. Burrows. Feeds on surface at night, leaves soil casts. Woodlands, pastures. T, ex Ic. [2]

Nocturnal Worm *Allolobophora nocturna*
LUMBRICIDAE BL 90–180 mm. Body cylindrical; dark reddish-brown; clitellum usually darker, on segments $\frac{1}{2}$27, 28–$\frac{1}{2}$35. Burrows. Feeds on surface at night, leaves soil casts. Pastures, occasionally arable land. Br, Fr, Ge. [3]

Pink Soil Worm *Allolobophora rosea*
LUMBRICIDAE BL 25–85 mm. Body cylindrical; pink or pinkish-grey, anterior segments reddish; clitellum usually orange, on segments 24, 25, 26–32, 33. Under stones, leaves, logs, in pastures, gardens, woodlands. Abundant. T. [4]

Tree Worm *Dendrobaena rubida*
LUMBRICIDAE BL 25–80 mm. Body cylindrical; dark red above, lighter red below; clitellum not prominent, on

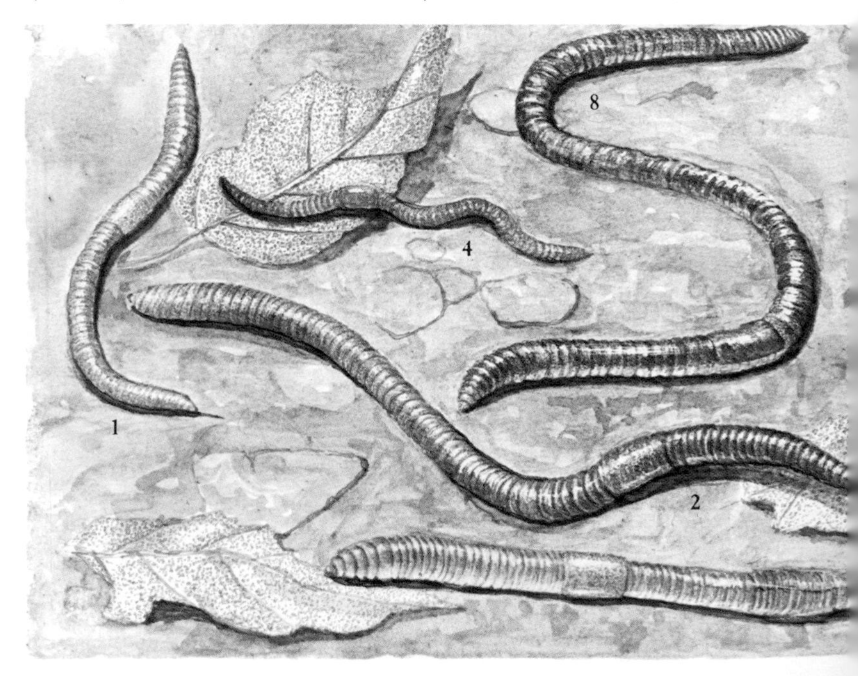

segments 26, 27–31, 32. Under stones, moss, in decaying tree stumps, often crawling on surface of soil or bark of trees at night. T, ex Ic. [**5**]

Venetian Worm *Dendrobaena veneta* LUMBRICIDAE BL 50–155 mm. Body cylindrical, flesh-pink with red or purple segmental bands above, clitellum on segments 24, 25, 26, 27–32, 33. In rich soil, esp under decaying vegetation. Br, Fr. [**6**] Brandling *Eisenia foetida* similar, but body yellower, clitellum on segments 24, 25, 26–32. T.

Chestnut Worm *Lumbricus castaneus* LUMBRICIDAE BL 30–70 mm. Body cylindrical, tail flattened; strongly iridescent chestnut or violet-brown above, brownish-yellow below; clitellum red or orange, on segments 28–33. Under stones, leaves, animal droppings. Moist, shady habitats in pastures, forests. T. [**7**] Lob or dew worm *Lumbricus terrestris* similar, lighter in colour, lacks iridescence; lives in burrow, esp in clay soils. T.

Rubescent Worm *Lumbricus festivus* LUMBRICIDAE BL 48–155 mm. Body cylindrical, tail slightly flattened; iridescent ruddy-brown above, pale yellow below; clitellum pink to reddish-brown, on segments $\frac{1}{2}$33, 34–39. Under stones, leaves, animal droppings, in pastures, on river banks. T, ex Ic, Fi. [**8**]

White Worm *Octolasion lacteum* LUMBRICIDAE BL 35–160 mm. Body cylindrical, slightly octagonal posteriorly; milk-white, grey, blue or rosy; clitellum pink or orange, on segments 30–35. In burrows, under leaf mould, in forests, gardens, cultivated fields. T, ex Ic. [**9**]

nematode *Mermis nigrescens* MERMITHIDAE BL ♀ $<$130 mm (♂ much smaller, rarely seen). Body threadlike, lacking segmentation, white. Larva parasitic in body cavity of insect or other invertebrate, adult free-living in soil or water. Usually seen after rain, when ♀ comes to surface and climbs plants to lay eggs. T. [♀ **10**] Many other spp of nematode live in soil.

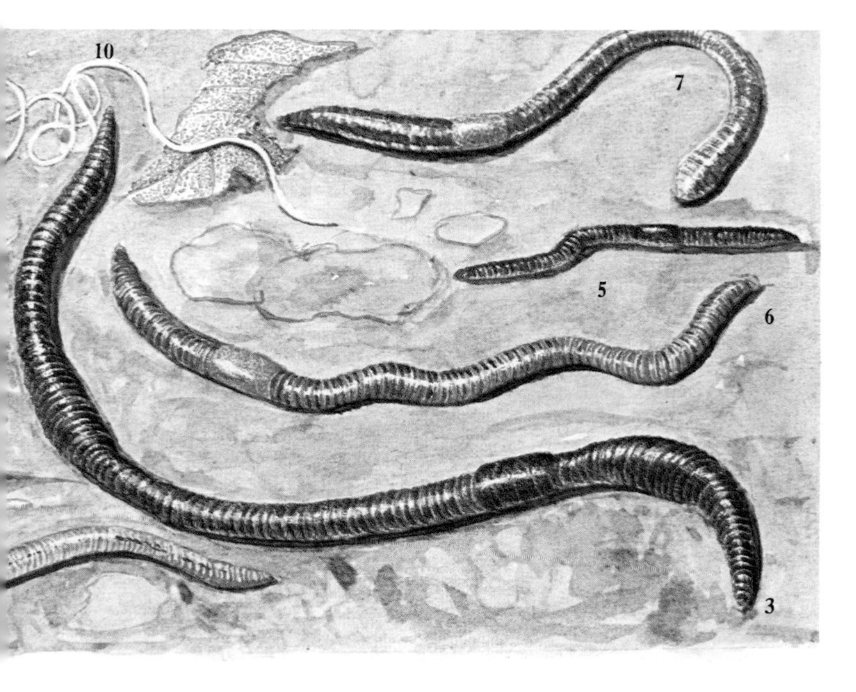

Millipedes have 2 pairs of legs to each body segment, centipedes have 1. Nocturnal; in daytime in litter, under loose bark, logs or stones, in soil. Centipedes feed on insects, spiders, small slugs; millipedes feed on decomposing vegetation, may damage plants.

centipede *Necrophloeophagus longicornis* GEOPHILIDAE BL 30–45 mm. Bright yellow, with short, yellow legs, dark red-brown head. Phosphoresces if disturbed. In soil, litter, under stones. T, ex Ic. [1] Several similar spp.

centipede *Lithobius duboscqui* LITHOBIIDAE BL 5·5–9·5 mm. Small, chestnut-brown; characterized by curling up when disturbed. Esp cultivated land, gardens, but also in woodland. T, ex Ic, Fi. [2] *Lithobius crassipes* L 8–11 mm, similar, in woodland, also curls up. T, ex Ic, Fi.

millipede *Polymicrodon polydesmoides* CRASPEDOSOMIDAE BL 17–21 mm. Grey-brown to dark brown; body of 28 or 30 segments with pronounced lateral expansions. Common in woodland litter, and on downs under stones, dead wood; frequent in British caves, where active all year. Br, Ir, Fr, No. [3] *Polydesmus angustus* similar, with 19 or 20 segments, in same habitats, also farmland, gardens. Br, Ir, Fr, Be, Ne, Ge.

Swift-footed Iulus *Tachypodoiulus niger* IULIDAE BL 19–50 mm. Brown-black or black with white legs; very active. Under surface debris in woodland, chalk grassland; often climbs trees, walls, at night, enters houses. Br, Ir, Fr, Be, Ne, Ge. [4]

Woodlice are the only crustaceans (crabs, shrimps) fully adapted to live on land. They remain vulnerable to desiccation, so seek out damp places. Some spp roll up to keep thin undersides moist, also as protection from predators. Feed on decaying plants or animals, moss, or bark; a few are carnivorous. ♀ carries eggs in a 'pouch' until miniature adults hatch.

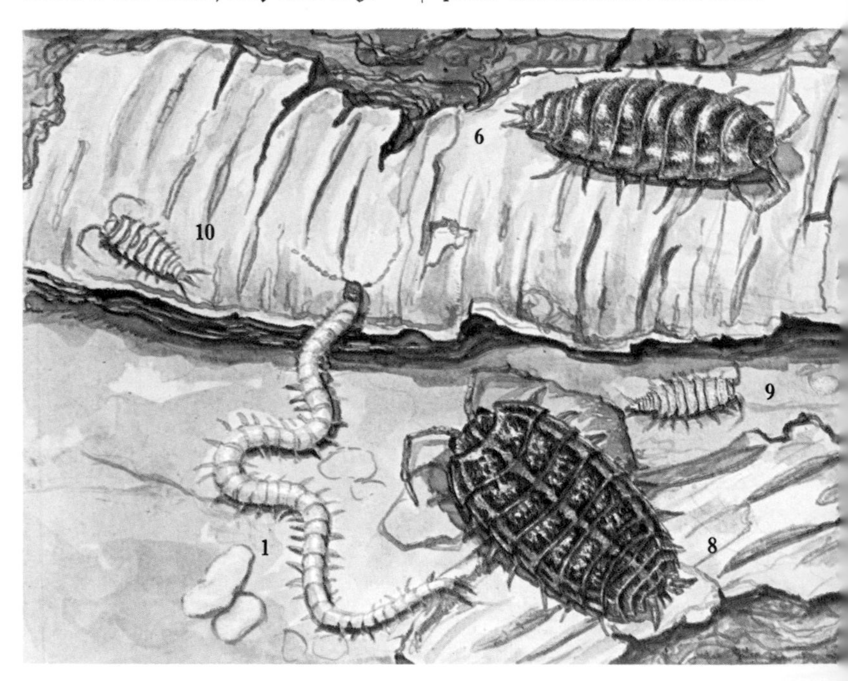

woodlouse *Philoscia muscorum* ONISCIDAE BL 6–12 mm. Body smooth, brownish, mottled, with dark median stripe; head contrasting black or dark brown; occasional yellow or red forms; abdomen distinctly narrower than thorax, legs elongate; active. Widespread, locally common, in grassland, hedgerows; less frequent in woodland. T, ex Ic. [5]

woodlouse *Cylisticus convexus* CYLISTICIDAE BL 10–16 mm. Body elongate, strongly convex, dark grey, occasionally brown; may roll into oval-shaped ball. Usually on chalk downs, sometimes in hedgerows, gardens, sea-shore. T. [6]

woodlouse *Platyarthrus hoffmannseggi* SQUAMIFERIDAE BL 3–4·5 mm. Body small, flattened, oval, tuberculate, white; appendages short, no eyes. In ants' nests, esp yellow ant *Lasius flavus* in Br, rarely outside. T, ex Ic, SC. [7]

woodlouse *Trachelipus rathkei* PORCELLIONIDAE BL 12–15 mm. Body surface granular, mottled brown and yellow with 3 pairs of longitudinal stripes; does not roll into ball. Widespread, locally common, in lowland habitats. T. [8]

woodlouse *Trichoniscus pygmaeus* TRICHONISCIDAE BL 2–2·5 mm (smallest sp in Br). Usually whitish or yellowish but with little pigmentation, occasionally brownish or pink; abdomen distinctly narrower than thorax. Widely distributed, probably common, in lowland habitats, typically associated with farming, horticulture. T. [9]

woodlouse *Haplophthalmus danicus* TRICHONISCIDAE BL 4 mm. Whitish or yellowish; body segments with longitudinal ridges and tubercles; unlike *T. pygmaeus*, margin of abdomen continuous with thorax. Widespread, usually on rotting tree-trunks. T. [10]

Such ubiquitous woodlice as *Porcellio scaber*, *Oniscus asellus* and *Trichoniscus pusillus* also in lowland fields and woods.

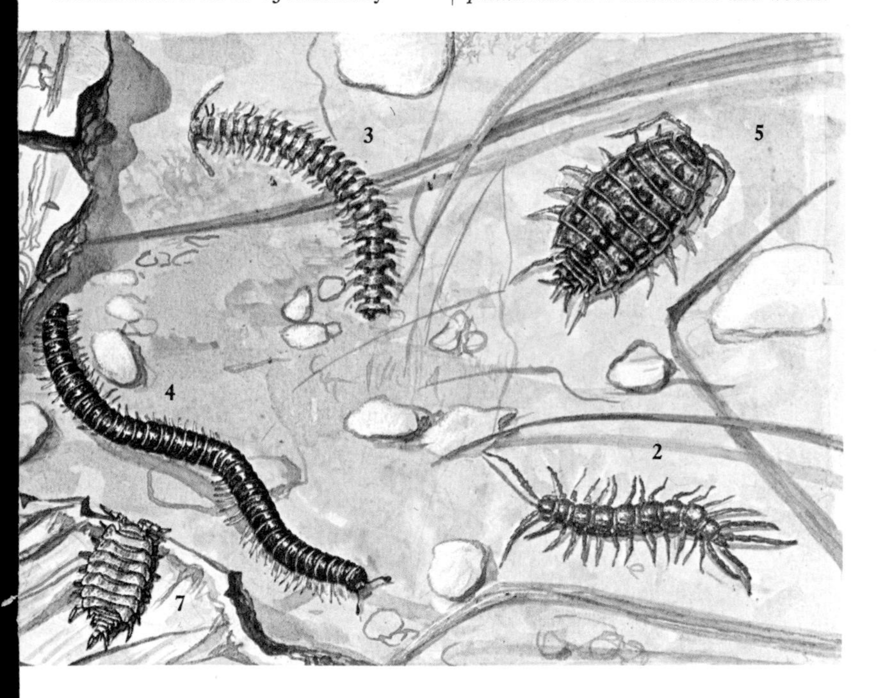

Dragonflies, damselflies (Odonata). Mostly large, with long, slender bodies, taking prey in flight. Metamorphosis incomplete; nymphs aquatic, predatory.

Common Coenagrion, Azure Damselfly *Coenagrion puella* COENAGRIIDAE BL 32–35 mm, WS *c*41 mm. ♀ yellow-green on sides, almost black on top of abdomen. Delicate, fluttery flight. Diurnal, May–Sep; common in grassland, esp near slow-flowing water. T, ex Ic. [♂ **1**]

Four-spotted Libellula *Libellula quadrimaculata* LIBELLULIDAE BL 45–48 mm, WS *c*75 mm. Body stout; wings each with 2 spots. Rapid flight. Diurnal, May–Aug; common in agricultural areas, esp near peat bogs. T, ex Ic. [**2**]

Emperor Dragonfly *Anax imperator* AESCHNIDAE BL 75–80 mm, WS *c*106 mm. ♀ with browner markings than ♂. Diurnal, but often flies at dusk, Jun–Sep; grasslands, woods, agricultural areas, esp near water. Powerful flier. ♀ lays eggs in waterplants just below surface. T, ex Ic, nFS, but rare Ir. [♂ **3**]

Cockroaches (Dictyoptera). 2 cerci, wings held flat over back. Omnivorous. Metamorphosis incomplete; nymphs wingless.

Dusky Cockroach *Ectobius lapponicus* BLATTIDAE BL ♂ 9–11 mm, ♀ 7–8 mm. Flattened, oval body. Diurnal, May–Oct; grasslands, hedgerows, woods. ♂ flies freely, often in trees or shrubs; ♀ probably non-flying, usually on ground. T, ex Ir, Ic, but only sBr. [**4**]

Crickets, grasshoppers (Orthoptera). Produce sounds by stridulation. Feed on plants, insects. Eggs laid in soil. Metamorphosis incomplete; nymphs like small, wingless adults.

Field Cricket *Gryllus campestris* GRYLLIDAE BL 17–24 mm, ♀ with sword-like ovipositor 9–11·5 mm. Head large, rounded, with long, slender antennae; hindwings reduced; 2 long cerci. Flightless. Nocturnal, May–Jul; well-drained grasslands, living in tunnel in ground. ♂ chirps loudly by rubbing forewings together in bursts, 1–6 (usually 3–4) per second. WE, CE. [♂ **5**]

Mole Cricket *Gryllotalpa gryllotalpa* GRYLLOTALPIDAE BL 30–50 mm. Brown, covered with velvety hair. Nocturnal, all year, flying early summer; cultivated areas, grasslands. Tunnels underground with specialized forelegs. May damage crops by eating roots. Nocturnal churring 'song' recalls nightjar. T, ex Ic, Fi, rare No. [♂ **6**]

Common Green Grasshopper *Omocestus viridulus* ACRIDIDAE BL 15–22 mm. Variable colour; wings reach tip of abdomen. Diurnal, Jun–Oct; in grasslands, woodland clearings. ♂ chirp is loud, 10–20 clicks per second for 10–20 seconds, ending abruptly. T. [♀ **7**]

Meadow Grasshopper *Chorthippus parallelus* ACRIDIDAE BL ♂ 10–16 mm, ♀ 16·5–22 mm. Hindwings usually reduced, forewings of ♂ reach tip of abdomen, but of ♀ only ½-way; colour variable. Largely diurnal, Jun–Oct; grasslands, cultivation. ♂ chirp lasts 1–3 seconds. T. [♀ **8**]

Great Green Bush-cricket *Tettigonia viridissima* TETTIGONIIDAE BL 41–54 mm, ♀ with long, slightly downcurved ovipositor 19–24 mm. No median keel on pronotum. Mainly diurnal, but activity extending into night, Jul–Oct; widespread but local in woodlands, hedgerows, shrubby areas. ♂ stridulation loud, harsh, high-pitched, continuing for minutes, even hours, with only brief pauses. T, ex Ic. [♀ **9**]

Wart-biter *Decticus verrucivorus* TETTIGONIIDAE BL 32–37 mm, ♀ with long, upcurved ovipositor 19–21 mm. Diurnal, Jul–Sep; open grasslands. ♂ stridulates in long bursts of 5 minutes of more, esp in warm sunshine, producing series of rapid clicks, like free-wheeling bicycle, gathering momentum to rate of 10 or more per second. T, ex Ic. [**10**]

3
8
2
7
9
10
4
1
6
5

springtails COLLEMBOLA BL 1–6 mm. Minute, grey or whitish; wingless, but many jump actively when disturbed using springing organ at rear. All year; under stones, in leaf-litter, in all habitats (many spp and huge numbers of individuals in most places). Feed on plant debris, fungi; sometimes pests of crops. T. [1]

barklouse *Amphigerontia bifasciata* PSOCOPTERA BL 4–5 mm. Delicate, pale green, with slender antennae, 2 pairs of transparent wings with darker veins. Spring–autumn; common in woodlands, hedgerows, esp lichen-covered branches, trunks. Feeds on algae, lichens, fungi, growing on bark. Nymph minute, wingless, in crevices in bark. T, ex Ic. [2] Many similar spp.

thrips, thunderflies THYSANOPTERA BL <1 mm (some <5 mm). Brownish or blackish; 2 pairs of wings, fringed with long hairs. Spring–summer; on fls, lvs, esp cereals. Many spp, some predatory, but mostly plant-feeding. Larvae feed on sap, causing small silvery spots on leaves. Adults swarm freely, esp in warm thundery weather: often 'fly in eye' is thrip. T. [3]

Lacewings (Neuroptera). Small to large; usually long antennae; soft, slender bodies; generally brown or green; 2 pairs of gauzy wings held roof-like over body at rest; slow fliers; biting mouthparts, feed on aphids, plant-lice, other insects. Metamorphosis complete.

green lacewing *Chrysopa carnea* CHRYSOPIDAE WS 27–30 mm. Delicate green with golden eyes; wings transparent, with green veins. Gentle, fluttering flight. Nocturnal, attracted to lights, but readily flies by day; all year, hibernating when cold (sometimes in houses); widespread, common in grasslands, woods, agricultural areas. Eggs laid on thin stalks on plants. Larva elongate; also predatory. T, ex Ic. [4] [larva **5**]

brown lacewing *Hemerobius humulinus* HEMEROBIIDAE WS 15–18 mm. Brown;

wings transparent, with brown veins and small spots along these. Slow, fluttery flight. Nocturnal, but will fly by day; Mar–Oct; widespread in woodland, hedgerows. Preys on aphids, scale insects. T, ex Ic. [6]

snakefly *Raphidia notata* MEGALOPTERA: RAPHIDIDAE BL *c*10 mm, WS *c*25 mm. Small head, front of thorax elongated like 'neck', ♀ has long ovipositor; shiny black with 2 pairs of transparent wings. Not very active. Diurnal, Jun–Aug; widespread, rarely abundant, in woodlands, hedgerows. Preys on other insects. Larva slender, thorax narrower than head or abdomen; also feeds on insects. T, ex Ic. [♀ 7]

scorpionfly *Panorpa germanica* MECOPTERA: PANORPIDAE WS 31–33 mm. ♂ has tip of abdomen with broad appendages curled up over body, ♀ has tip pointed; transparent wings roughly equal in size, marked with black. Slow flight. Diurnal, May–Jun; widespread, not numerous, in woodlands, hedgerows. Feeds on live and dead insects. Larva caterpillar-like; lives in galleries in soil. T, ex Ic. [♂ 8]

Earwigs (Dermaptera) have short, hardened forewings (elytra); ear-shaped hindwings; stout pincers at rear.

Lesser Earwig *Labia minor* FORFICULIDAE BL 5–6 mm. Reddish-brown with black head, pincers at tip of abdomen (♂ has bases of pincers slightly separated and straight, ♀ closer together and more curved). Mostly diurnal, summer; widespread, not abundant, in grasslands, downs, wood edges. Flies freely; attracted to rotting vegetation, dung. Metamorphosis incomplete; young like small, wingless adult. T, ex Ic. [♀ 9]

Common Earwig *Forficula auricularia* FORFICULIDAE BL 11–14 mm. Dark brown with yellowish elytra, prominent pincers (in ♀ straight, in ♂ strongly curved). Nocturnal, hiding by day; summer; widespread, common under loose bark, logs. T. [♂ 10]

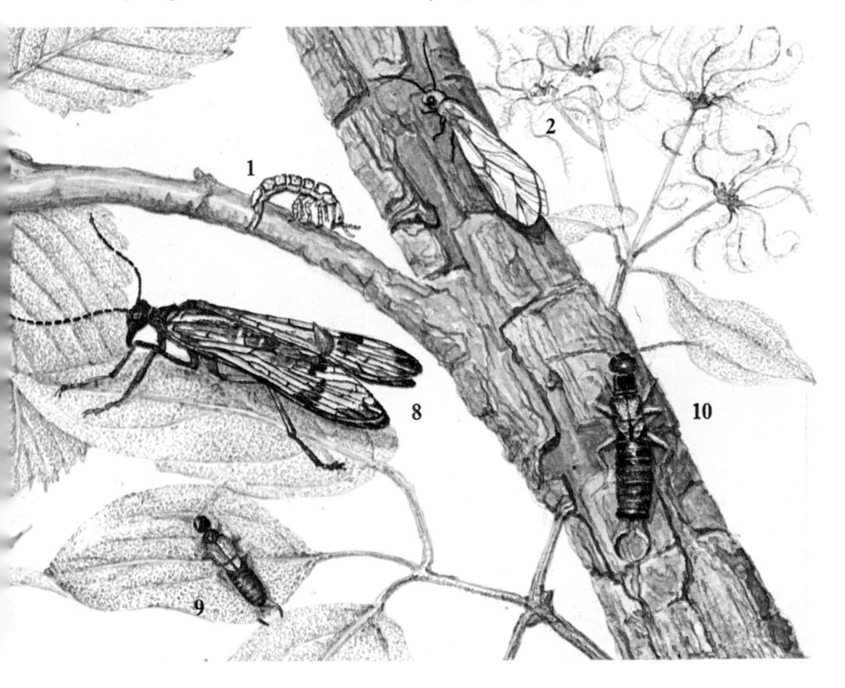

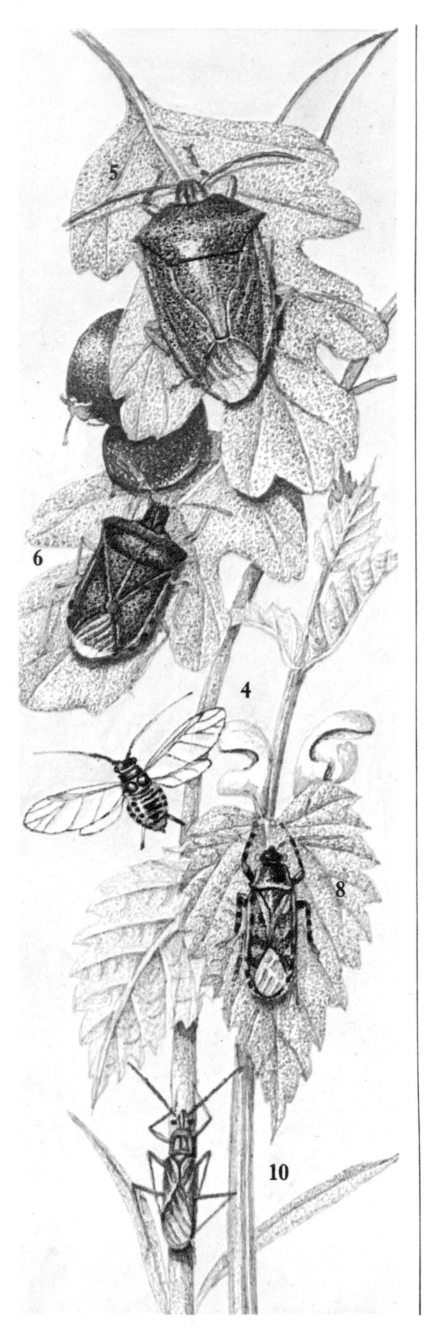

Spittle-bugs, leafhoppers, plantlice, aphids, cicadas, land bugs, water bugs (Hemiptera). Minute to large, with sucking mouthparts for extracting liquids from plants, animals. 2 pairs of wings held roof-like over body (Homoptera) or flat (Heteroptera). Forewings (hemelytra) have basal part hardened, rest membranous. Metamorphosis incomplete; nymph often wingless.

froghopper *Neophilaenus lineatus* CERCOPIDAE BL 4·5–6·5 mm. Colour variable, but generally brown with yellow streak along margin of forewings. Leaps actively if disturbed. Summer; widespread, common in grasslands. Feeds on plants. Nymph produces froth ('cuckoo spit') on plant and lives inside this, protected by it from desiccation and predators. T, ex Ic. [**1**] [cuckoo spit **2**]

leafhopper *Cicadula sexnotata* CICADELLIDAE BL 3–3·5 mm. Slender, yellow, with 2 black spots on thorax, small spots on head. Diurnal, summer; widespread, abundant in cultivation, grasslands, sometimes in large numbers on barley, oats, sugar-beet. T. [**3**]

Potato or **Peach Aphid** *Myzus persicae* APHIDIDAE BL 2 mm. ♂ with 2 pairs of transparent wings; ♀ sometimes wingless, with 2 long cornicles (tubes) on abdomen; antennae long; shiny green or yellowish. All year; abundant in agricultural land, hedgerows. Pest of potatoes, other crops. Eggs overwinter on peach. ♀ also viviparous. T. [winged ♀ **4**] Many spp of aphids, including greenfly and blackfly, common on crops, other plants; often transmit virus diseases. Some spp excrete sweet 'honeydew'.

Hawthorn Shield-bug *Acanthosoma haemorrhoidale* ACANTHOSOMATIDAE BL 13–15 mm. 1st segment of antennae reaches beyond apex of head, side of thorax pointed. Aug–Apr, adults overwinter; common in mixed woods, hedgerows. Feeds on hawthorn lvs, berries, also oak lvs. T, ex Ic. [**5**]

forest-bug *Pentatoma rufipes* PENTATOMIDAE BL 11–13 mm. Short blunt projection on each side of thorax; head, thorax brown or red-brown, densely speckled with black; marks on closed wings form red-brown spot. Jul–Aug; common in woodlands. Feeds on plants including oak, cherry. T, ex Ic. [6]

Pied Shield-bug *Sehirus bicolor* CYDNIDAE BL 5·5–7·5 mm. Head, thorax black and grey; large dark patch, sometimes L-shaped, across middle of each forewing. 2 broods, Apr–Jul, Aug–Apr, overwintering; common in hedgerows, pastures, other grasslands. Flies freely, feeds on many trees in spring. T, ex Ir, Ic. [7]

Nettle Ground-bug *Heterogaster urticae* LYGAEIDAE BL 6–7 mm. General appearance grey-black; head black, with long hairs, thorax black, paler on posterior margin; dark marks on forewings; 3 black stripes on legs. Sep–Jun, overwintering; numerous where common nettles grow. T, ex Ic, No. [8]

European Chinch-bug *Ischnodemus sabuleti sabuleti* LYGAEIDAE BL 4–5 mm. Slender, with parallel sides; head, thorax, abdomen black; short- and long-winged forms. 2 broods, Apr–early Jul, Aug–Apr, overwintering as adult and nymph; in damp meadows, pastures. T, ex Ic. [9]

stilt-bug *Berytinus minor* BERYTINIDAE BL 5–7 mm. Head shorter than pronotum, which has prominent keel; body, forewings yellow-brown; winged and wingless forms. Aug–May, overwintering; common on downs, drier areas. Feeds on aphids, plants. T, ex Ic. [10]

Creeping Thistle Lace-bug *Tingis amplicata* TINGIDAE BL 3–4 mm. Broad shape, expanded thorax; antennae stout, 4 segments; reticulations (network of intersecting lines) give whole body lace-like appearance, enhanced by grey-brown covering of powdery wax. Aug–Jun, overwintering; common in hedgerows, meadows. Feeds on creeping thistle. sBr, Fr, Ne, Ge, Cz, No. [11]

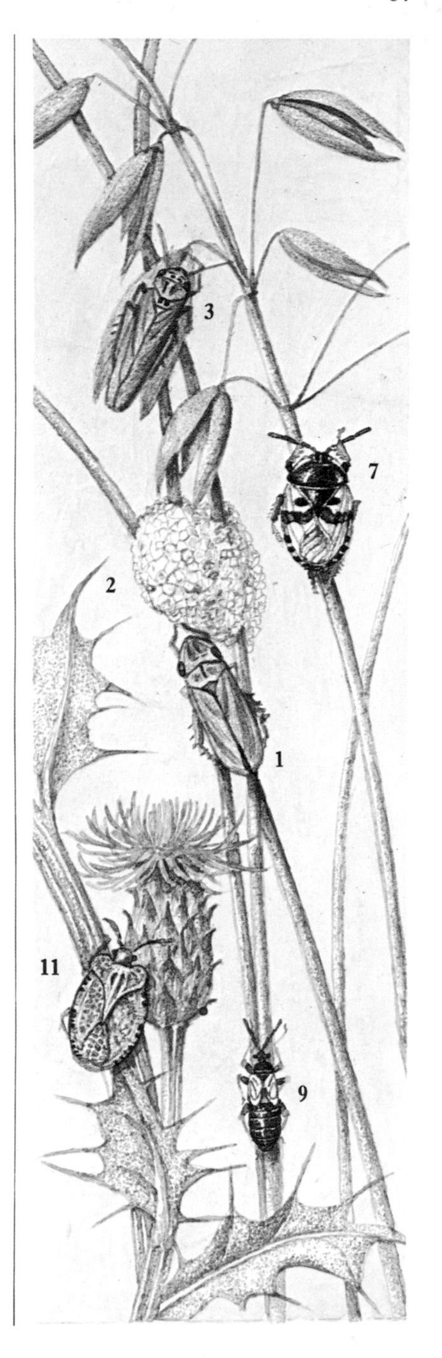

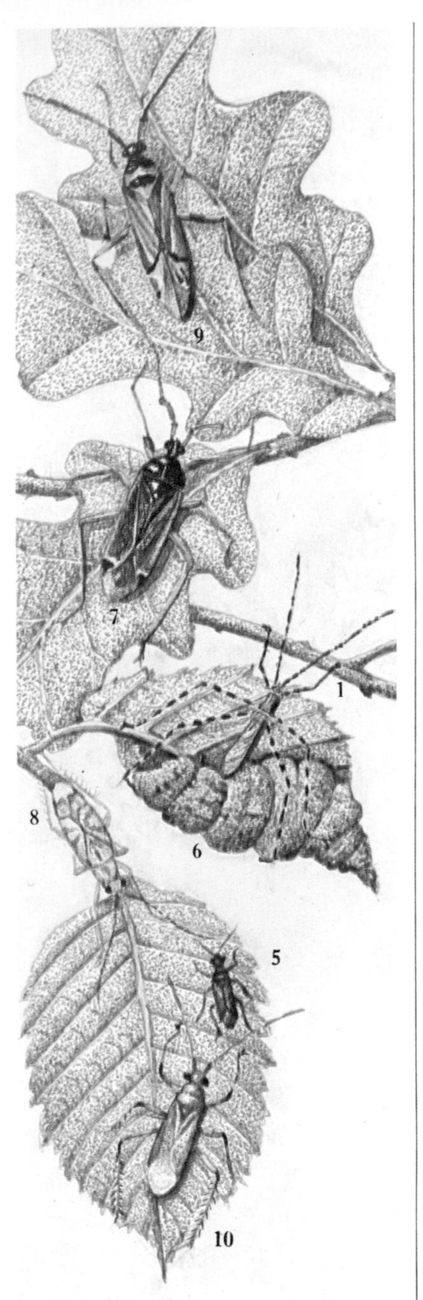

Thread-legged Bug *Empicoris vagabundus* REDUVIIDAE BL 6–7 mm. Antennae longer than body; rostrum strongly curved. Jul–May, overwintering; common in mixed woodlands. Feeds on aphids, other insects. Eggs laid on branches, trunks of trees. T, ex Ic. [**1**]

Field Damsel-bug *Nabis ferus* NABIDAE BL 8–9 mm. Pale grey, with black lines along thorax; forewings well developed, densely hairy. Aug–Jun, overwintering; widespread in cornfields, meadows. Preys on other insects. T, ex Ic. [**2**]

Common Damsel-bug *Nabis rugosus* NABIDAE BL 6·5–7·5 mm. Forewings yellow-brown or grey-brown, darker than *N. ferus*; thorax more heavily marked; anterior legs with black spots (unmarked in *N. ferus*). Aug–Jul, overwintering; common in meadows, cultivation. Preys on other insects. T, ex Ic. [**3**]

Ant Damsel-bug *Aptus mirmicoides* NABIDAE BL 7–8 mm. Antennae shorter than body, wings usually shorter than abdomen (but sometimes full-winged). More active at night, Jul–May, overwintering; common in low herbage in meadows, downlands, drier areas. Preys on other insects. T, ex nBr, Ic, Fi. [**4**]

Elm Gall-bug *Anthocoris gallarum-ulmi* ANTHOCORIDAE BL 4–5 mm. Head black, 1st antennal segment black, 2nd pale. Probably 2 generations, Aug–Jun, overwintering; common in woods, hedgerows, in leaves of elm curled by aphid *Eriosoma ulmi*, feeding on the aphids. T, ex Ic. [**5**] [curled leaf **6**]

oak-bug *Harpocera thoracia* MIRIDAE BL 6–7 mm. Knob on 2nd antennal segment of ♂. May–Jun; common in oak woods. Feeds on oaks, flies freely. Eggs laid in May, hatch following spring. T, ex Ic, Fi. [♂ **7**]

Delicate Apple Capsid *Malacoris chlorizans* MIRIDAE BL 3·5–4·5 mm. Pale green with darker green marks. 2 generations, Jun–Aug; widespread, common in woods, hedgerows. Mainly predatory, but also feeds on elm, apple, hazel. Eggs overwinter. T, ex Ic. [8]

oak-bug *Cyllecoris histrionicus* MIRIDAE BL 6–8 mm. Head and front of thorax black. Jun–Sep; common in oak woods. Mostly predatory on other insects. T, ex Ic. [9]

Black-kneed Capsid *Blepharidopterus angulatus* MIRIDAE BL 5–6 mm. Green; antennae, rear edge of pronotum, and bases of tibiae ('knees') black. Jun–Oct; widespread, common on many trees in woods, hedgerows. Predatory. Eggs laid on young twigs, may cause some distortion. T, ex Ic. [10]

European Tarnished Plant-bug *Lygus rugulipennis* MIRIDAE BL 5–6 mm. Grey with black markings; densely covered with fine hairs. 2 generations, Jul, Sep–Mar, overwintering; common in hedgerows, pastures, cultivation. Feeds on plants, esp Chenopodiaceae, clovers, sometimes pest of potatoes. T, ex Ic. [11]

capsid-bug *Notostira elongata* MIRIDAE BL 6–8·5 mm. Head with longitudinal groove between eyes, basal segments of antennae and tibiae covered with long hairs; ♂ has grey-green margins to wings and thorax, black central area, grey-green legs; overwintering ♀ brown with shorter wings, summer ♀ green. 2 broods, Apr–Aug, Sep; common, widespread in hedgerows, pastures, meadows. Feeds on grasses. T, ex Ic, Cz, No. [♂ 12]

Meadow Plant-bug *Leptopterna dolobrata* MIRIDAE BL ♂ 8–9 mm, ♀ 7·5–9·5 mm. ♂ full-winged, ♀ usually short-winged; ♀ grey-green, with black marks along body. Jun–Sep; common, widespread in grasslands. Feeds on grasses; can be pest in pastures, in Sweden also on cereals (damage appears as spotting on lvs, deformation of fls). T, ex Ic. [♂ 13]

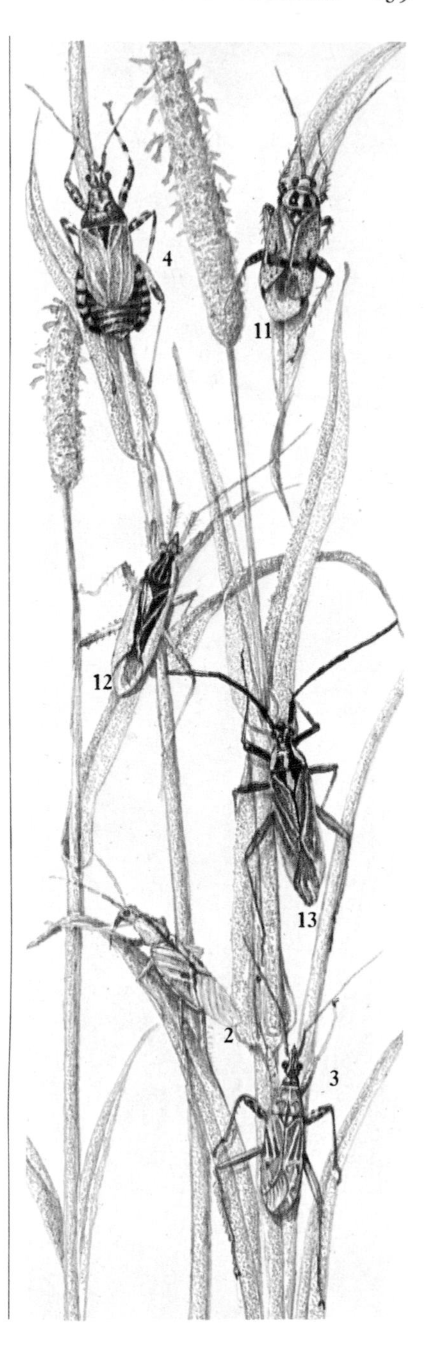

Butterflies, moths (Lepidoptera). Small to large, covered by scales giving colour and pattern; usually proboscis for sucking nectar. Butterflies mostly day-fliers, with clubbed antennae; chrysalis without cocoon. Moths mostly night-fliers, with thin, pointed or pectinate (feathery) antennae; cocoon or chamber formed round pupa. Metamorphosis complete.

Scarce Swallowtail *Iphiclides podalirius* PAPILIONIDAE WS 65–85 mm. Pale lemon with black stripes. Often 2 broods, Mar–Sep; meadows, pastures, esp near orchards. Caterpillar feeds on sloe, fruit trees. Fr, Lu, Be, Ge, Cz, Po, occasional Ne, De, sSw, vagrant Br. [1]

Black-veined White *Aporia crataegi* PIERIDAE WS 57–66 mm. White, with black body and wing-veins. May–Jul; meadows, agricultural land, esp orchards. Caterpillar feeds on hawthorn, fruit trees. T, ex Br (extinct), Ir, Ic, nFS. [U **2**]

Small White *Pieris rapae* PIERIDAE WS 46–54 mm. Forewings white with faint black apex, ♀ also with 2 black spots on upperside and underside, ♂ with 2 spots on underside only, faintly visible on upperside; hindwings faintly yellowish on upperside with small black mark at front edge. Apr–Oct; common, widespread, esp meadows and cultivated land. Caterpillar feeds on various Cruciferae, can be pest of *Brassica*. T. [U **3**]

Large White *Pieris brassicae* PIERIDAE WS 56–66 mm. Similar to *P. rapae*, but larger with more conspicuous spots on forewings. 2 broods, Apr–Jun, Jul–Oct; widespread, often abundant in agricultural areas, grassland; migratory. Caterpillar feeds gregariously on Cruciferae, esp *Brassica*; can be pest. T.

Green-veined White *Pieris napi* PIERIDAE WS 36–44 mm. Underside of hindwings with yellowish-green veins. T, ex Ic.

Orange-tip *Anthocharis cardamines* PIERIDAE WS 38–48 mm. ♂ white with bright orange tip to forewings, which ♀ lacks. Apr–Jun; common in hedgerows, damp meadows. Caterpillar feeds on Cruciferae. T, ex Ic. [♂ U **4**]

Clouded Yellow *Colias croceus* PIERIDAE WS 46–58 mm. Uppersides with black outer edges; ♀ with orange spot in middle of hindwings. Rapid flier, migrant. May–Sep; meadows. Caterpillar feeds on Leguminosae. T, ex Ic, but migrant Br, Ir, Ne, sFi, Sw, sNo. [U **5**] Several broadly similar spp.

Brimstone *Gonepteryx rhamni* PIERIDAE WS 52–60 mm. ♀ as ♂ but pale yellow-green. Rapid flier. Mar–Jun, Sep (but adults overwinter, may appear on any mild winter day); common open woodlands, hedgerows. Caterpillar feeds on buckthorn. T, ex Ic, far north. [♂ U **6**]

Wood White *Leptidea sinapis* PIERIDAE WS 37–50 mm. Underside of hindwings grey and white. Most active in sunshine. 2 broods, Apr–May, Jul–Sep; open woodlands, wood margins. Caterpillar feeds on Leguminosae. T, ex Ic, far north. [**7**]

Purple Emperor *Apatura iris* NYMPHALIDAE WS 62–80 mm. ♀ grey-brown with larger white markings than ♂, no purple gloss. Powerful, rapid flier, keeping near tree-tops. Jul–Aug; woodlands, esp of oaks with sallows, attracted to carrion. Caterpillar feeds on willows. sBr, Fr, Lu, Be, Ge, Cz, Po, sFi. [♂ **8**]

White Admiral *Ladoga camilla* NYMPHALIDAE WS 52–62 mm. Underside of forewings red-brown with white marks, hindwings with grey base and 2 rows of dark spots. Flapping flight, often glides. Jun–Jul; woods. Caterpillar feeds on honeysuckle. Br, Fr, Lu, Be, De, Ge, Cz, Po, sSw, sFi. [**9**] Several rather similar spp.

Poplar Admiral *Ladoga populi* NYMPHALIDAE WS 70–82 mm. Powerful flier, often glides. Jun–Jul; locally common in woodlands. Caterpillar feeds on poplars. Fr, Lu, Be, De, Ge, Cz, sFS, ?Po. [**10**]

8
10
6
2
9
3
5
7
4
1

Large Tortoiseshell *Nymphalis polychloros* NYMPHALIDAE WS 50–64 mm. Distinguished from small tortoiseshell by reddish-black basal area on hindwings, larger size. 1 brood, Jul–May, adult overwinters; local in open woods. Caterpillar feeds on elms, willows, other trees. Br, Fr, Be, Ne, Ge, Cz, Po, sFS. [1]

Peacock *Inachis io* NYMPHALIDAE WS 54–58 mm. Underside dark black-brown. Rapid flier. Jul-Mar, adult overwinters; widespread in hedgerows, grasslands. Attracted to flowers in autumn. Caterpillar feeds clustered in webs on nettles. T, ex Ic, nFS. [2]

Red Admiral *Vanessa atalanta* NYMPHALIDAE WS 56–63 mm. Forewings velvet-black with red band in middle, white patches near apex. Rapid flight; migratory. May–Oct, including immigrants from S Europe; widespread in meadows, hedgerows, downlands. Attracted to flowers in autumn. Caterpillar solitary, feeds on nettles, living in curled leaf held with silk. T, ex Ic, nFS. [3]

Painted Lady *Cynthia cardui* NYMPHALIDAE 54–58 mm. Underside of hindwings whitish-brown with blue-centred spots near margin. Rapid flier, migrant, unable to survive winter in N Europe; May–Sep; widespread in meadows, hedgerows. Caterpillar feeds on thistles, nettles, other plants. T, ex Ic, but rare in north. [4]

Small Tortoiseshell *Aglais urticae* NYMPHALIDAE WS 44–50 mm. Hindwings have whole basal half black-brown. Rapid flier. 2 broods, Jun–Aug, Sep, then overwintering; common in meadows, hedgerows, farmland. Attracted to flowers. Caterpillar feeds in clusters on nettles. T, ex Ic. [5]

Comma *Polygonia c-album* NYMPHALIDAE WS 44–48 mm. Both wings orange-brown with blackish patches. 2 broods, Jun–Jul, Aug, then overwintering; widespread, in meadows, hedgerows, farmland. Feeds at flowers. Caterpillar feeds on nettles, hops, willows, other trees. T, ex nBr, Ir, Ic, nFS. [6]

Map Butterfly *Araschnia levana* NYMPHALIDAE WS 33–38 mm. 1st brood wings orange-brown with pale-lined black bases, irregular spots and yellow patches; 2nd brood blackish-brown with yellow or white bands, and reddish lines near outer margins. 2–3 broods, Apr–Sep; common in open woodlands. Caterpillar feeds on nettles. Fr, Lu, Be, sDe, Ge, Cz, Po, (Br, but extinct). [2nd brood **7**]

Silver-washed Fritillary *Argynnis paphia* NYMPHALIDAE WS 54–72 mm. Rapid flight. Jun–Aug; locally common, in open woodlands. Caterpillar feeds mainly on violets. T, ex nBr, Ic, nFS. [8]

High Brown Fritillary *Argynnis adippe* NYMPHALIDAE WS 51–63 mm. Both wings bright orange-brown with black markings. Rapid flight. Jun–Jul; widespread, in open woods, meadows. Caterpillar feeds on violets. T, ex nBr, Ir, Ic, nFS. [9]

Small Pearl-bordered Fritillary *Boloria selene* NYMPHALIDAE WS 37–43 mm. Both wings orange-brown with black markings, hindwings usually having round black spot near base. Low, rapid, gliding flight. Jun–Jul; widespread, common in woodland clearings, wood edges, grasslands. Caterpillar feeds on violets. T, ex Ir, Ic. [U **10**]

Pearl-bordered Fritillary *Boloria euphrosyne* NYMPHALIDAE WS 37–48 mm. Similar to *B. selene*. Jun–Jul; widespread, locally common in open woodlands, hedgerows, meadows. Caterpillar feeds on violets. T, ex Ic, local Ir. [U **11**]

Glanville Fritillary *Melitaea cinxia* NYMPHALIDAE WS 33–42 mm. Both wings orange-brown with brown crosslines, white patches along outer edges, 4 or 5 black spots towards outer margin of hindwings. May–Jun; local in meadows, woodland edges. Caterpillar feeds on plantains, also hawkweeds. Fr, Lu, Be, Ne, De, Ge, Cz, Po, sFS, 1 area in sBr. [U **12**]

2
1
6
3
7
5
4
9
8
10
11
12

Marbled White *Melanargia galathea* SATYRIDAE WS 48–54 mm. Both wings white to cream with black markings. Jun–Jul; locally common in meadows, hedgerows, wood edges. Caterpillar feeds on grasses. sBr, Fr, Be, Lu, Ge, Cz, Po. [1]

Woodland Ringlet *Erebia medusa* SATYRIDAE WS 43–50 mm. Spots near outer margins, surrounded by orange-yellow in ♀. May–Jun; meadows, open woods. Caterpillar feeds on grasses. eFr, Be, Ge, Cz, Po. [♂ **2**] Several similar spp.

Meadow Brown *Maniola jurtina* SATYRIDAE WS 45–52 mm. ♂ dark brown with paler areas (orange in ♀); underside of hindwings with 2 black dots in outer part. Jun–Aug; very common, in meadows, downs, hedgerows. Caterpillar feeds on grasses. T, ex Ic, nFS. [♀ **3**] [caterpillar **4**]

Ringlet *Aphantopus hyperanthus* SATYRIDAE WS 42–50 mm. ♂ almost black with pale fringes, usually 2 obscure spots near apices of both wings; ♀ paler brown with better-defined spots, usually white-centred on hindwings. Jun–Jul; common in open woodlands, meadows, downs. Caterpillar feeds on grasses. T, ex nBr, Ic, nFS. [U **5**]

Gatekeeper, Hedge Brown *Pyronia tithonus* SATYRIDAE WS 34–39 mm, ♂ smaller than ♀. ♂ underside of forewings similar to upper side, of hindwings mottled grey-brown to reddish with yellow-brown band; ♀ brighter orange-yellow. Jul–Aug; locally common in hedgerows, rough pastures, woodland rides. Often at bramble flowers. Caterpillar feeds on grasses. sBr, sIr, Fr, Lu, Be, Ne, De, Ge, Cz, sPo. [♂ **6**]

Speckled Wood *Pararge aegeria* SATYRIDAE WS 37–45 mm. Underside of forewings similar to upperside but paler, of hindwings yellow-brown with small, white spots near outer margin. 2 broods, Apr–May, Aug–Oct; common in woodlands, mostly in shade. Caterpillar feeds on grasses. T, ex Ic, nFS. [**7**]

Woodland Brown *Lopinga achine* SATYRIDAE WS 52–65 mm. Both wings grey-brown with series of large, yellow-ringed, black spots near slightly wavy outer margins. Jun–Jul; locally common in woodlands, staying in shade. Caterpillar feeds on grasses. Fr, Lu, Be, Ge, sFi, sSw, ?Cz, ?Po. [U **8**]

Duke of Burgundy Fritillary *Hamearis lucina* NEMEOBIIDAE WS 29–37 mm. Undersides orange-brown with dark markings similar to uppersides; 2 bands of yellow-white patches across hindwings. May–Jun; locally common in woods, sunny glades; feeds at sap-flows on trees. Caterpillar feeds on cowslips, primroses. sBr, Fr, Lu, Be, Ge, Cz, Po, sSw. [**9**]

Brown Hairstreak *Thecla betulae* LYCAENIDAE WS 35–37 mm, ♂ smaller than ♀. Forewings brown with orange band (small, sometimes absent in ♂), hindwings with 2 orange marks near short 'tail'. Jul–Aug; local in open woods, flying high among trees. Caterpillar feeds on sloe, plum, birches. sBr, sIr, Fr, Lu, Be, Ne, De, Ge, Cz, Po, sFS. [♀ **10**]

Purple Hairstreak *Quercusia quercus* LYCAENIDAE WS 25–29 mm. ♀ wings black with bright bluish only towards base of forewings; undersides grey with white transverse line, of forewings with yellow-orange marks, of hindwings with orange-ringed black spot and 2nd orange mark near short 'tail'. Jul–Aug; local in woods, esp of oaks, flies near tree tops. Caterpillar feeds on oaks. T, ex nBr, nIr, Ic, but only sFS. [♂ **11**] [caterpillar **12**]

Ilex Hairstreak *Nordmannia ilicis* LYCAENIDAE WS 33–38 mm. Forewings dark brown, with or without orange patch; hindwings also dark brown, with red marks near short 'tail'. Jun–Jul; locally common in oak woods, esp on hill slopes. Caterpillar feeds on various oaks. Fr, Lu, Be, Ne, De, Ge, Cz, Po, sSw. [U **13**]

12
10
13
11
1
3
4
5
8
6
9
7
2

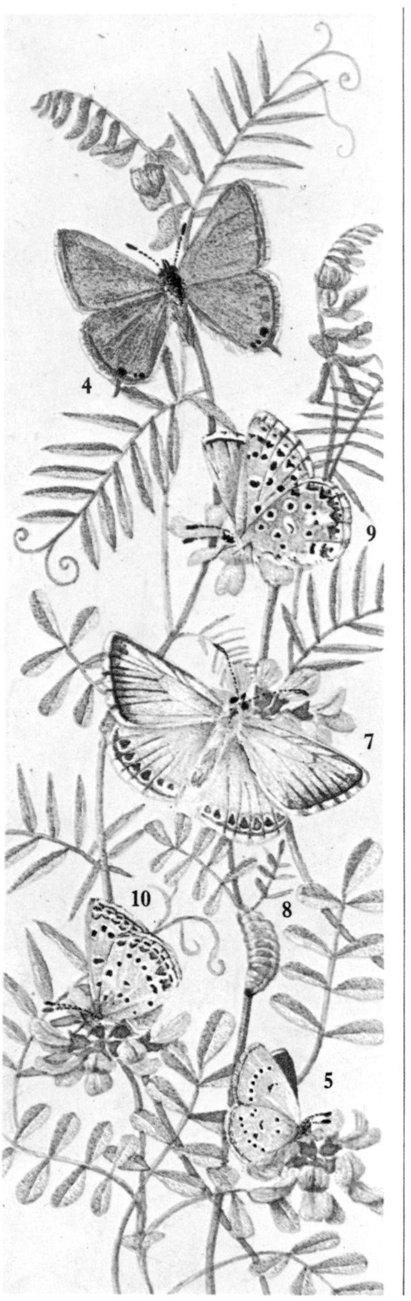

Small Copper *Lycaena phlaeas* LYCAENIDAE WS 25–32 mm. Underside of forewings pale orange with grey-brown border, yellow-ringed black spots, of hindwings grey-brown. Actively dashes from flower to flower; sits with wings spread. Several broods, Mar–Sep; widespread in meadows, pastures, downs, hedgerows. Caterpillar feeds on docks, sorrels. T, ex Ic. [**1**] [caterpillar **2**]

Sooty Copper *Heodes tityrus* LYCAENIDAE WS 29–33 mm. ♂ wings dark grey-brown with obscure darker spots, narrow orange band near outer margins; ♀ forewings orange with brown edges, black patches, hindwings dark brown, with black spots in bold orange band near outer margin. 2 broods, Apr–May, Aug–Sep, locally common in meadows, pastures, usually in lowlands. Caterpillar feeds on docks. Fr, Lu, Be, Ne, sDe, Ge, Cz, Po. [**3**]

Long-tailed Blue *Lampides boeticus* LYCAENIDAE WS 31–37 mm. ♀ browner than ♂, violet-blue confined to basal halves; undersides of ♂ and ♀ grey-brown with whitish and dark grey stripes, hindwings also with 2 black and orange 'eyes' at rear. Several broods, Jun–Sep; widespread in meadows, rough pastures; migratory. Caterpillar feeds on Leguminosae. sBr, Fr, Lu, Be, Ne, Ge, Cz, Po. [♂ **4**]

Small Blue *Cupido minimus* LYCAENIDAE WS 21–25 mm. Both wings dark brown, of ♂ with scattering of blue scales. Often 2 broods, Apr–Sep; local, but generally in colonies, in grassy meadows, often in limestone areas. Caterpillar feeds on Leguminosae. T, ex nBr, nIr, Ic, nFS, rare Ne. [**5**]

Baton Blue *Philotes baton* LYCAENIDAE WS 21–25 mm. ♀ blackish with some blue near bases; undersides of ♂ and ♀ grey, heavily spotted with black, hindwings with orange band near outer margin. 2 broods, Apr–Jun, Jul–Sep; widespread in meadows, pastures. Caterpillar feeds on thymes. Fr, Be, Ge, Cz, Po, sFi. [♂ **6**]

Chalk-hill Blue *Lysandra coridon* LYCAENIDAE WS 31–37 mm. ♀ brown with variable blue flush; underside of forewings of ♂ and ♀ grey with black spots, white-edged, of hindwings browner with white-ringed black spots, paler orange marks near outer margin, white wedge in middle, blue-green wash at base. Jul–Aug; locally common in limestone areas, chalk downs. Caterpillar feeds on vetches, trefoils. sBr, Fr, Be, Ne, Ge, Cz, Po. [♂ 7] [caterpillar 8]

Adonis Blue *Lysandra bellargus* LYCAENIDAE WS 29–37 mm. ♀ brown with variable blue wash, orange spots on outer edges; underside of ♀ similar to ♂ but darker brown with less or no blue at base. 2 broods, May–Jun, Jul–Aug; local in limestone areas, chalk downs. Caterpillar feeds on Leguminosae. sBr, Fr, Lu, Be, Ge, Cz, Po. [♂ 9]

Common Blue *Polyommatus icarus* LYCAENIDAE WS 29–37 mm. ♂ pale purplish-blue, with fine black outer margins; ♀ brown, usually suffused blue, with well-marked line of orange spots near outer edges; underside of ♀ browner than ♂, with stronger markings. 2–3 broods, Apr–Sep; common in meadows, grasslands, agricultural areas. Caterpillar feeds on trefoils, vetches, clovers. T, ex Ic. [♂ U 10]

Dingy Skipper *Erynnis tages* HESPERIIDAE WS 27–29 mm. Underside pale brown with white dots along margins and centre of hindwings. Rapid, whirring flight, low among grass. 2 broods, May–Jun, Jul–Aug; widespread in meadows, rough grasslands, downs. Caterpillar feeds on Leguminosae. T, ex Ic, Fi, but local Ir, only sSC. [11]

Chequered Skipper *Carterocephalus palaemon* HESPERIIDAE WS 27–29 mm. Dark brown with large, yellow-orange patches. Rapid, darting flight. Jun–Jul; local in woods, esp of oak, forest edges, in small colonies. Caterpillar feeds on grasses, esp bromes. T, ex Ir, Ic, Ne, but local Br, De, SC. [U 12]

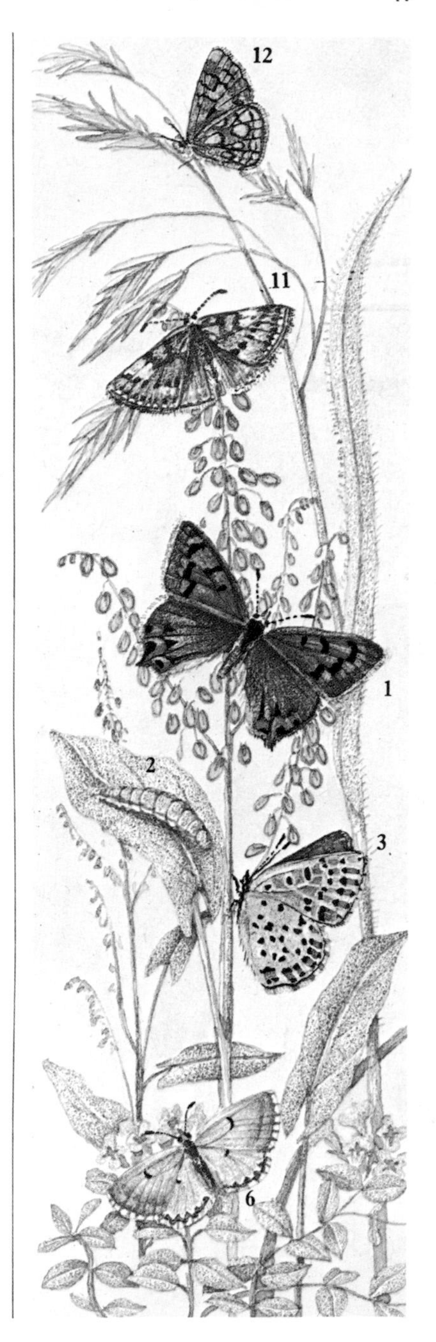

Privet Hawk-moth *Sphinx ligustri* SPHINGIDAE WS 90–110 mm. Hindwings pale, often pink, with 3 blackish bands; abdomen with pink and blackish bands, brown stripe down middle. Rapid flight; camouflaged at rest when forewings cover pink areas. Nocturnal, attracted to light, Jun–Jul; locally common, rarely abundant, in woods, hedgerows. Caterpillar feeds on privet, ash, lilac, other shrubs; pupates underground. T, ex Ic. [1]

Death's-head Hawk-moth *Acherontia atropos* SPHINGIDAE WS 102–135 mm. Hindwings yellow with 2 dark brown bands. Squeaks through proboscis, *eg* when handled. Nocturnal, May–Oct; locally common in farmlands, but migratory. Feeds on honey inside bee-hives, also on sap from tree wounds. Caterpillar feeds esp on potato, can be serious pest in CE, also on bittersweet, snowberry; pupates underground. T, ex Ic. [2]

Yellow-tail, Gold-tail *Euproctis similis* LYMANTRIIDAE WS 32–45 mm. ♂ has pectinate (feathery) antennae, 1–2 black spots at rear of forewings, small yellow tuft; ♀ has simple antennae, larger orange-yellow tuft. Nocturnal, often attracted to light, Jun–Jul; widespread, often abundant in woodlands, hedgerows. Caterpillar feeds on hawthorns, oaks, beech, fruit trees, often spinning extensive webs. T, ex Ic, ?No. [♀ 3]

Cinnabar *Tyria jacobaeae* ARCTIIDAE WS 35–44 mm. Hindwings crimson, with narrow, black fringe. Nocturnal, attracted to light; also flies by day, May–Jun; common in grasslands, downs, hedgerows. Caterpillar feeds communally on ragwort, Jul–Aug; pupates underground. T, but ?Ic. [4] [caterpillar 5]

Peach-blossom *Thyatira batis* THYATIRIDAE WS 35–38 mm. Hindwings brownish-white, base white on ♂, pale brown on ♀. Nocturnal, Aug–Sep; widespread, sometimes common, in woods, hedgerows. Caterpillar feeds on brambles. T, ex Ic. [6]

Red Underwing *Catocala nupta* NOCTUIDAE WS 70–80 mm. Forewings pale grey; hindwings crimson, with black bands. Camouflaged at rest when forewings cover crimson. Nocturnal, attracted to light, Aug–Sep; sometimes abundant in woods, hedgerows. Caterpillar feeds at night on poplars, willows. T, ex Ic. [7] Several related spp with red hindwings.

Flounced Rustic *Luperina testacea* NOCTUIDAE WS 30–33 mm. Forewings, red-brown or grey-brown with obscure markings; hindwings white with dark marginal line. Rapid flight. Nocturnal, attracted to light, Aug–Sep; common, in grasslands, agricultural areas. Caterpillar feeds on grass roots. T, ex Ic, Fi. [8] Many rather similar spp.

Plain Golden-Y *Autographa jota* NOCTUIDAE WS 34–42 mm. Forewings marbled brown, with small white Y- or V-shaped mark above dot; hindwings smoky grey-brown. Rapid flight. Nocturnal, attracted to light, May–Jul; common in woods, hedgerows, grasslands. Caterpillar feeds on dead-nettles, nettles, woundworts, hawthorns. T, ex Ic. [9] Several similar spp.

Large Emerald *Geometra papilionaria* GEOMETRIDAE WS 45–56 mm. Both wings green, with white lines. Fluttery flight. Nocturnal, attracted to light, Jun–Jul; common in hedgerows, woods. Caterpillar feeds on birches, hazel, beech, moving in looping manner. T, ex Ic. [10] Several similar spp.

Chalk Carpet *Scotopteryx bipunctaria* GEOMETRIDAE WS 28–33 mm. Forewings grey-brown on white, with darker wavy crosslines and black colon in middle; hindwings paler smoky-grey, with obscure darker bands. Fluttery flight. Nocturnal, but easily disturbed by day, Jul–Aug; in chalk grasslands, on limestone. Caterpillar feeds on clovers, trefoils, moving in looping manner. Br, Fr, Ge, Cz, Po. [11]

10
1
7
3
6
9
5
2
4
11
8

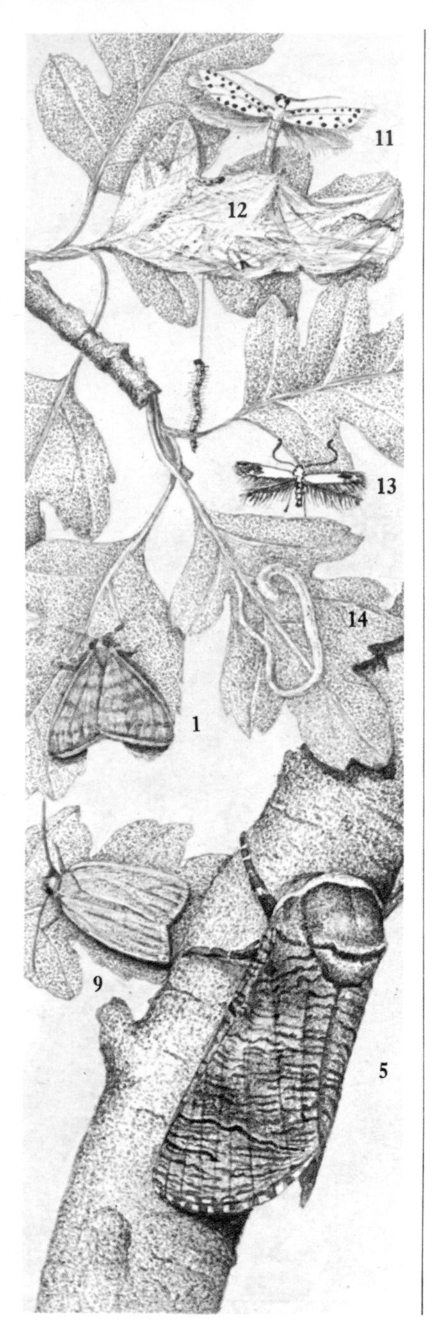

Winter Moth *Operophtera brumata* GEOMETRIDAE WS ♂ 25–28 mm, ♀ virtually wingless. Hindwings grey-brown. Nocturnal, Oct–Feb; common in woods. agricultural lands, orchards. ♀ climbs trees to lay eggs; ♂ flies on milder winter nights. Caterpillar feeds on hawthorns, oaks, fruit trees, Apr–May; may cause serious defoliation. T. [♂ **1**]

Six-spot Burnet *Zygaena filipendulae* ZYGAENIDAE WS 28–35 mm. Hindwings deep crimson with narrow, blue-black fringe. Whirring flight. Diurnal, esp active in bright sunshine, Jun–Jul; widespread, common in meadows, grasslands, downs. Caterpillar feeds on Leguminosae; makes conspicuous yellow cocoon on grass. T, ex Ic. [**2**] [caterpillar **3**] [pupa **4**] Several rather similar spp.

Goat Moth *Cossus cossus* COSSIDAE WS 70–94 mm, ♂ smaller than ♀. Body fat, brown; wings rather broad. Nocturnal, May–Jul; widespread in woodlands. Caterpillar bores into ash, elms, birches, willows, for 3–4 years, before pupating in ground. Caterpillar and its tunnels smell of goats. T, ex Ic. [**5**]

Crimson and Gold *Pyrausta purpuralis* PYRALIDAE WS 14–20 mm. Hindwings with black edge, yellowish band near outer margin, paler base. Mainly diurnal, active in sunshine, but will fly at night. 2 broods, May–Jun, Jul–Aug; pastures, grasslands, woodland edges. Caterpillar feeds on mint, thyme, spinning lvs together. T, ex Ic. [**6**]

grass-moth *Chrysoteuchia culmella* PYRALIDAE WS 18–22 mm. Long palps form protruding 'snout' in front of head; hindwings grey-brown. Rests along grass stems with wings rolled round body. Crepuscular, but will fly by day if disturbed, Jun–Jul; common, often abundant, in grasslands. Caterpillar feeds on roots of grasses, hibernates; sometimes pest, esp on golf courses, pastures. T, ex Ic. [**7**] Several similar spp.

Large White Plume-moth *Pterophorus pentadactyla* PTEROPHORIDAE WS 26–29 mm. Legs very long; forewings divided into 2 lobes, hindwings into 3 lobes. Fluttering flight; rests with wings rolled at right-angles to body, looking T-shaped. Crepuscular, attracted to light, but easily disturbed by day. Often 2 broods, Jun–Aug; common in hedgerows, woodlands. Caterpillar feeds on bindweeds. T, ex Ic. [8] Several related spp.

Green Tortrix *Tortrix viridana* TORTRICIDAE WS 17–24 mm. Hindwings smoky grey-brown. Crepuscular, but easily disturbed by day, Jun–Jul; common in woods, esp of oaks, also hedgerows. Caterpillar rolls lvs of oaks, other trees, herbaceous plants; may cause serious defoliation. T, ex Ic. [9]

long-horned moth *Adela reamurella* ADELIDAE WS 14–16 mm. Antennae of ♂ extremely long, slender, of ♀ shorter; forewing metallic green; hindwings dark purplish-brown. Flutters in large groups, with antennae outstretched, by trees, shrubs, settling readily on lvs. Diurnal, May–Jun; common, sometimes abundant, in hedgerows, wood edges. Caterpillar in leaf-litter. T, ex Ic, Fi, No. [**10**]

Small Ermine, Common Hawthorn Ermel *Yponomeuta padella* YPONOMEUTIDAE WS 16–24 mm. Forewings white, dusted with black scales. Crepuscular, Jul–Aug; common, widespread in hedgerows, woodlands. Caterpillar gregarious, spins webs; colonies may cover large areas of hedges or trees with silk; sometimes pest of fruit trees. T, ex Ic. [11] [webbing **12**] Several similar spp.

leaf-miner *Lyonetia clerkella* LYONETIDAE WS 8–9 mm. Small, inconspicuous. 2 broods, Jun–Aug; common in woods, hedgerows. Caterpillar makes winding mine inside leaf of apple, hawthorn, other tree; later emerges to pupate on underside of leaf. T, ex Ic. [13] [leaf-mine **14**] Several related spp.

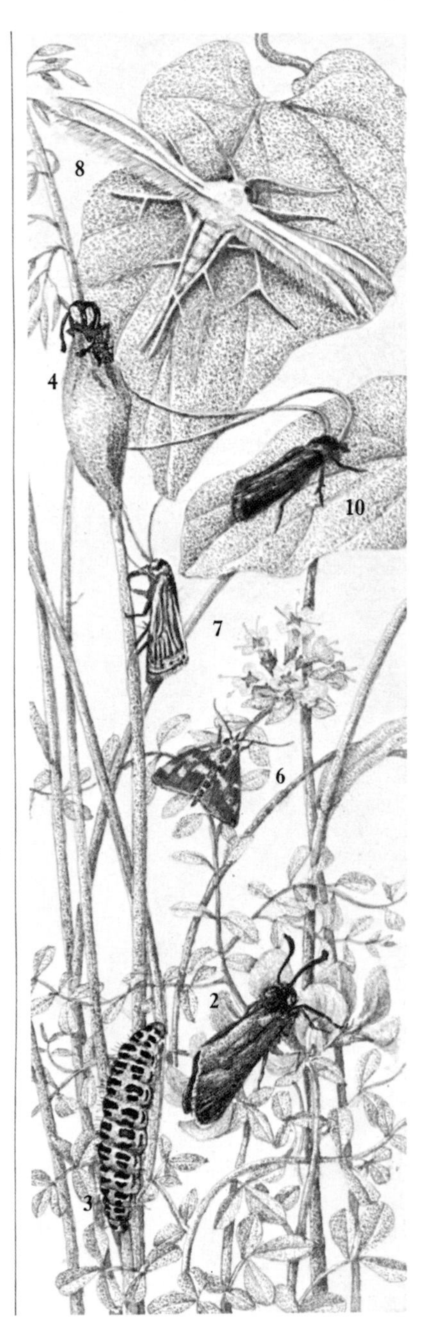

Flies (Diptera). Tiny to large, slender to stout, usually large-headed with prominent eyes; only 1 pair of transparent wings; hind pair reduced to knob-like halteres (balancers); mouthparts for sucking liquids, often modified for piercing. Metamorphosis complete. Larvae, or maggots, legless.

crane-fly, daddy-long-legs *Ctenophora atrata* TIPULIDAE BL 12–21 mm, WS 25–36 mm. Body slender, wings outspread at rest; ♀ black or yellow-and-black. Slow flier; hangs from vegetation by long legs. Nocturnal, summer; widespread in woodlands. Larva ('leatherjacket') lives in rotting wood. T, ex Ic. [♂ **1**] Many other spp; leatherjackets may become agricultural pests.

winter gnats TRICHOCERIDAE BL <7 mm, WS <16 mm. Like small, slender crane-flies; distinguished by short, curved lower vein in wings. Do not bite man. Diurnal, most of year, but more obvious in dancing swarms on mild winter days; widespread, common in most habitats. Larva in leaf-mould. T, ex Ic. [**2**]

mosquito *Aedes rusticus* CULICIDAE WS 6–7 mm. Slender; thorax with black stripes; abdomen with median stripe formed by dark grey patch on each segment. 2 broods, Apr–May, Jul–Aug; widespread in woods, hedgerows. ♀ bites, extracts blood. Larva aquatic. T, ex Ic, ?Fi, but only sSC. [**3**] Many other spp.

fungus-gnat *Dynatosoma fuscicornis* MYCETOPHILIDAE BL *c*8 mm. Delicate, thorax humped, antennae long and thread-like; wings with smoky patches. ♂s in swarms under trees, except in cold weather. Diurnal, mainly spring–autumn; common, widespread in woods, hedgerows. Larva feeds in fungi. T, ex Ic. [**4**]

St Mark's Fly *Bibio marci* BIBIONIDAE BL 10–11 mm. Antennae short, below eyes; foretibiae with large projections; body shiny black, hairy. Rather sluggish in flight. Diurnal, spring; common in woods, grasslands, downs. Larva in leaf-mould. T, ex Ic. [**5**]

black-fly *Simulium ornatum* SIMULIIDAE BL 2–4 mm. Humpbacked, antennae with 11 segments; all black except for silvery 2nd abdominal segment, yellowish legs. Emerges in large swarms. Diurnal, several broods, Feb–Apr, Jun–Nov; very common, in grasslands, farmlands, esp near water. ♀ bites; can transmit parasite to cattle. Larva aquatic. T. [**6**]

wheat midge *Contarinia tritici* CECIDOMYIIDAE BL *c*2 mm. Thorax and abdomen golden-yellow; wings iridescent, hairy; Diurnal, Jun; widespread in farmlands. Eggs laid on wheat. Larva feeds in wheat ovary, pupates in soil; can cause serious losses. T, ex Ic. [**7**]

cleg, dun, horse-fly *Haematopota pluvialis* TABANIDAE BL 8–10 mm. Slim, wings sloped roof-like over body at rest; tuft of long, black hairs covers most of head; dull-coloured, but eyes iridescent, purplish. Diurnal, May–Sep; widespread in farmland, esp near livestock. ♀ sucks blood; alights silently; bite causes painful swelling. Larva carnivorous; lives in wet soil. T, ex Ic. [**8**]

horse-fly, thunder-fly *Chrysops caecutiens* TABANIDAE BL 7–10 mm, WS 19–21 mm. Eyes iridescent green; wings of ♂ brown, of ♀ with brownish patches. Diurnal, Jun–Aug; common in woods, farmland, esp near water. ♀ sucks blood, of man and cattle; flies with loud hum, but settles silently, gives painful bite. Larva aquatic. T, ex Ic. [♀ **9**]

stout, horse-fly *Tabanus sudeticus* TABANIDAE BL 24–25 mm, WS 48–50 mm. Stoutly built, antennae rather scimitar-shaped; ♀ has short, stout proboscis. Diurnal, Jun–Aug; common in woods, farmland. ♀ sucks blood of cattle and horses; will bite man, causing painful swelling. ♂ feeds on nectar. Both make loud hum in flight. Larva lives in moist places. T, ex Ic. [**10**]

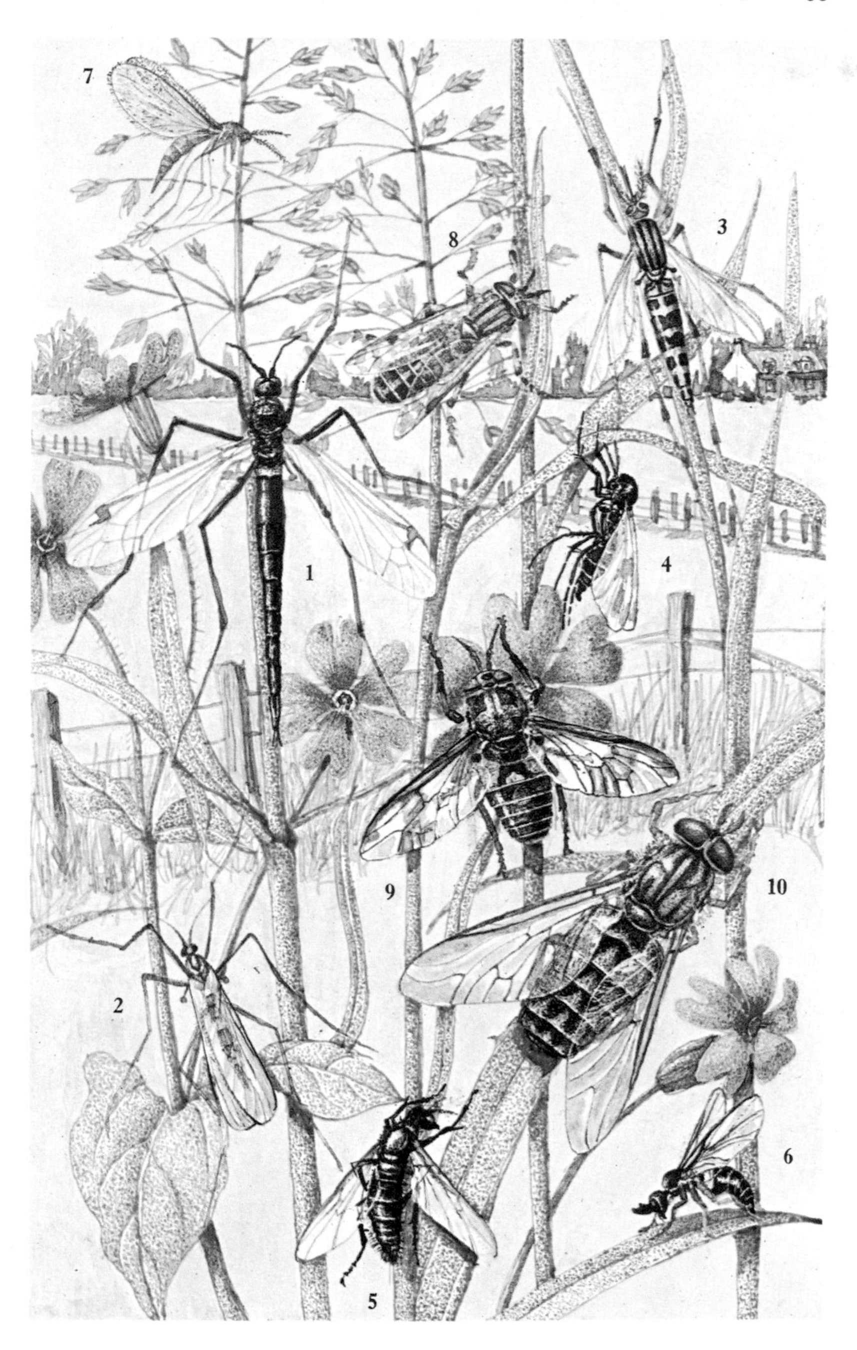
7
8
3
1
4
9
10
2
6
5

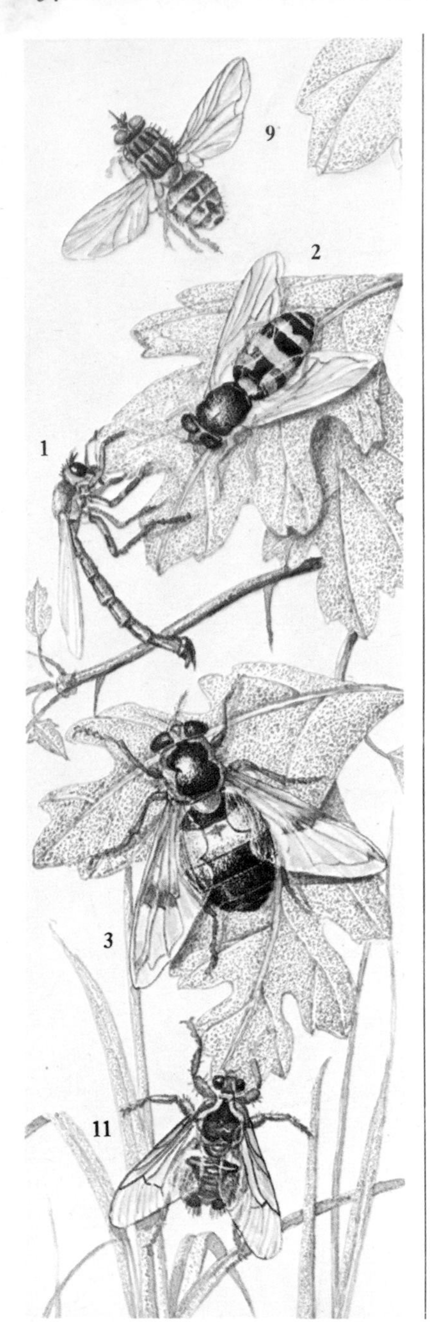

robber-fly *Leptogaster cylindrica* ASILIDAE BL 9–11 mm. Head with prominent beard; body very slender, brown with dark stripe on abdomen; legs stout, with dark stripe on hind pair. Diurnal, May–Aug; local in meadows, hedgerows, woods. Catches small flies in flight. Larva in leaf-litter. Br, CE, NE. [1]

hover-fly *Syrphus vitipennis* SYRPHIDAE BL 11–13 mm, WS *c*25 mm. Thorax greenish, yellow at rear; abdomen banded. Hovers over fls in sunshine. Diurnal, spring–summer; common, widespread in woods, hedgerows. Larva maggot-like; feeds on aphids. T, ex Ic. [2]

hover-fly *Volucella pellucens* SYRPHIDAE BL 12–16 mm. Shiny black, with white band across abdomen; wings with dark spots, orange at base. Active, flies rapidly, hovers freely over fls. Diurnal, Apr–Jul; widespread in woods, hedgerows. Larva scavenges in wasps' nest. T, ex Ic. [3]

fly *Nemorilla floralis* TACHINIDAE BL 6–7 mm. Black with golden sheen; abdomen particularly reflective. Diurnal, summer; common in woods, hedgerows. ♀ seeks out caterpillar on which to lay egg; larva becomes internal parasite. T, ex Ic. [4]

bluebottle *Calliphora vicina* CALLIPHORIDAE BL 10–12 mm, WS 21–23 mm. Metallic blue-grey, hairy. Diurnal, summer; common in hedgerows, farmland, woods. Flies with loud hum. Attracted to dead animals, dung; liable to transmit diseases from refuse to human food. ♀ commonly enters houses; ♂ tends to stay nearer flowers. Larva ('maggot') in decaying animal matter, excrement. Fishermen use maggots as bait. T.
[5] Greenbottle *Lucilia caesar*, similar but metallic green; common, habits and habitats similar. T. Several similar spp to both.

Cluster-fly *Pollenia rudis* CALLIPHORIDAE BL 9–10 mm, WS 18–22 mm. Golden, hairy; abdomen with black bands. Diurnal, May–Oct, overwinters; common in woods, hedgerows. Sometimes congregates to overwinter. Larva believed parasitic on earthworms. T, ex Ic. [6]

warble, gad-fly *Hypoderma bovis* OESTRIDAE BL 13–15 mm. Rather bee-like, hairy. Diurnal, May–Jun; common in farmland, pastures, usually near cattle. Larva burrows under skin, then moves round towards animal's back, producing swellings ('warbles'); eventually breaks out through skin to pupate in ground. Cowhides spoilt by warble-fly damage. T, ex Ic. [7]

Beet-fly, Mangold-fly *Pegomyia hyoscyami* MUSCIDAE BL 5–6 mm. Head grey, thorax pale brown; abdomen swollen at apex. Diurnal, usually several overlapping broods, Apr–Sep; common in agricultural land. 2 races: pale form on henbane; darker on beet, mangolds. Larva lives in lvs, produces winding leaf-mine, later large blisters; can be serious pest. T. [8]

Stable-fly, Biting House-fly *Stomoxys calcitrans* MUSCIDAE BL 6–8 mm. Hairy, with short but prominent proboscis protruding from head; abdomen greyish. Diurnal, May–Oct; widespread in agricultural lands, hedgerows, esp common near farms, stables. Sucks blood, biting livestock and man. Larva lives in stable-litter, straw. T, ex Ic. [9]

Sweat-fly *Hydrotaea irritans* MUSCIDAE BL 4–7 mm. Wings blackish with yellow base, halteres yellow. Diurnal, Jul–Aug; widespread in woods, rough pastures, hedgerows, esp common near bracken. Does not bite, but very irritating: swarms round horses, cattle, attracted by secretions, esp wounds; round human beings, attracted by perspiration. Larval habit not known. T, ex Ic. [10]

Forest-fly *Hippobosca equina* HIPPOBOSCIDAE BL 9–10 mm. Forelegs broad, jutting out; movements crab-like, clinging; whole body flattened. Diurnal, May–Oct; common in hedgerows, grasslands. Parasitic, living on bodies of horses, cattle; will bite man. Single larva develops inside ♀. T, ex Ic. [11]

Bees, wasps, ants, ichneumons, sawflies (Hymenoptera). Usually 2 pairs of wings, but many wingless; mouthparts for biting (wasps, ants, sawflies, ichneumons) or sucking (bees). Except sawflies (sub-order Symphata), all have narrow 'wasp-waists' (sub-order Apocrita). Many Apocrita live socially in large colonies. Metamorphosis complete. Larvae legless maggots; those of sawflies caterpillar-like.

Wheatstem-borer, Corn Sawfly *Cephus pygmaeus* CEPHIDAE BL 8–10 mm. No waist, bright yellow legs. Diurnal, Jun–Jul; common in farmlands. Eggs laid in stems of cereals, also meadow-grasses. Larva burrows in stem, and breaks top off; overwinters in basal part. T, ex Ic. [1]

Large Red-tailed Bumblebee *Bombus lapidarius* BOMBIDAE BL ♂ 15–18 mm, ♀ 24–27 mm, ☿ 12–18 mm. Orange-red on last 3 abdominal segments. Diurnal, Apr–Aug; widespread in grasslands, hedgerows, cultivation. ♀ emerges early, feeds esp at dead-nettle, bugle; ♂ later, at thistles and others. Nest colony underground, sometimes in hole in wall. T, ex Ic. [♀ 2]

Common Garden Bumblebee, carder-bee *Bombus pascuorum* BOMBIDAE BL ♂ 15–18 mm, ♀ 18–22 mm, ☿ 12–15 mm. Generally light brown. Diurnal, May–Aug; widespread in hedgerows, agricultural lands, pastures. ♀ emerges earlier than ♂. Nest colony above ground, *eg* in old birds' nest, wall. T, ex Ic. [♀ 3]

Moss Bumblebee, carder-bee *Bombus muscorum* BOMBIDAE BL ♂ 11–13 mm, ♀ 19–22 mm, ☿ 15–17 mm. Diurnal, May–Aug; widespread in hedgerows, pastures. Colony frequently above ground; nest constructed by plaiting grass or moss with legs and jaws (hence 'carder-bee'). T, ex Ic, but only sFS. [♀ 4]

Knapweed Bumblebee, carder-bee *Bombus sylvarum* BOMBIDAE BL ♂ 15–18 mm, ♀ 12–22 mm, ☿ 10–16 mm. Abdomen with grey and white bands, apex red-brown on ♂, brown on ♀. Diurnal, May–Aug; common in grasslands, downs. ♀ emerges in early spring. Later, frequent at thistles, knapweeds, scabious. Colony usually above ground, sometimes in old birds' nest. WE, CE, sFS. [♀ 5]

Common Wasp *Vespula vulgaris* VESPIDAE BL ♂ 13–17 mm, ♀ 15–20 mm, ☿ 11–14 mm. Unlike bees, wasps hold wings lengthwise along body at rest; ♀s and ☿s sting. Face with black anchor-mark. Diurnal, May–Oct; esp in autumn, ubiquitous. ♀ emerges early in spring, starts nest in hole in ground, which grows as more and more ☿s mature; ♂s hatch Aug–Oct. Nest made of wood-pulp. Feeds on fruit juices; kills insects to feed larvae. T, ex Ic. [☿ 6] Several similar spp.

Red Wasp *Vespula rufa* VESPIDAE BL ♂ 13–16 mm, ♀ 15–20 mm, ☿ 10–12 mm. Markings reddish; hairy. Diurnal, May–Aug; widespread in hedgerows, pastures, but rarely enters houses. Nest colony in hole in ground. Habits similar to other wasps. WE, CE, sFS. [♀ 7]

Hornet *Vespa crabro* VESPIDAE BL ♂ 20–24 mm, ♀ 26–32 mm, ☿ 19–23 mm. ♀ and ☿ sting severely. Diurnal, summer; local in woodlands. Nest colony above ground, sometimes in hollow tree or bird nestbox; nest looks like irregularly ridged papier-mâché ball. WE, CE, sFS. [☿ 8]

Hairy-legged Mining-bee *Dasypoda plumipes* MELITTIDAE BL 12–15 mm, ♂ smaller than ♀. Hindlegs of ♀ with large, hairy pollen-basket, of ♂ less hairy. Diurnal, summer; common in hedgerows, meadows. Feeds at flowers of Compositae. Nest chamber at end of tunnel, in sandy soil; often in colonies. WE, CE, sFS. [♀ 9]

Common Leafcutter-bee *Megachile centuncularis* MEGACHILIDAE BL 11–12 mm. ♀ with pollen-brush on underside. Diurnal, Jun–Aug; common in hedgerows, woodlands. Nest in crevice. With her jaws, ♀ cuts out disc-shaped pieces of lvs, carries to nest, rolls into cylinders, and fills with pollen, inserting 1 egg in each. WE, NE. [♀ 10]

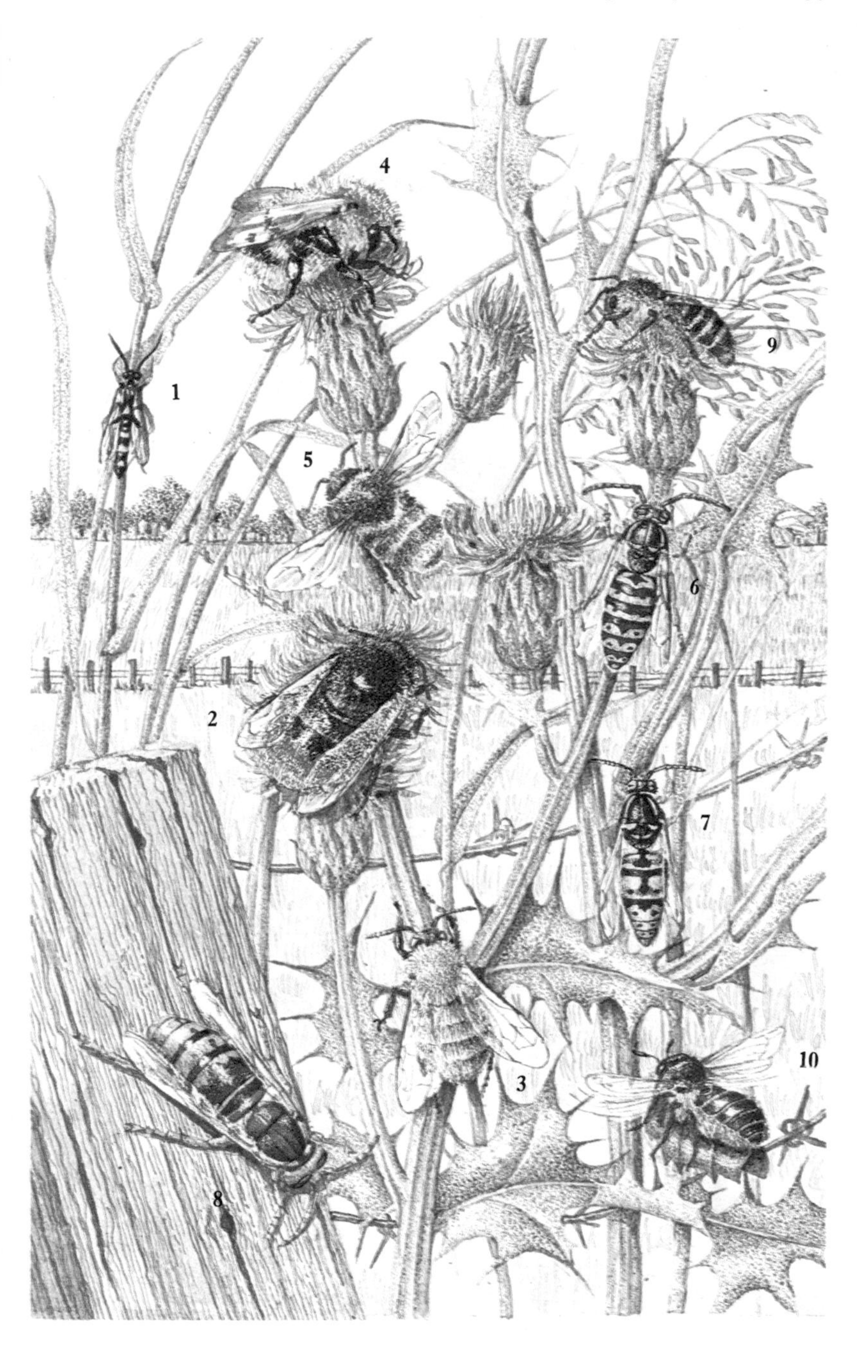
4
9
1
5
6
2
7
10
3
8

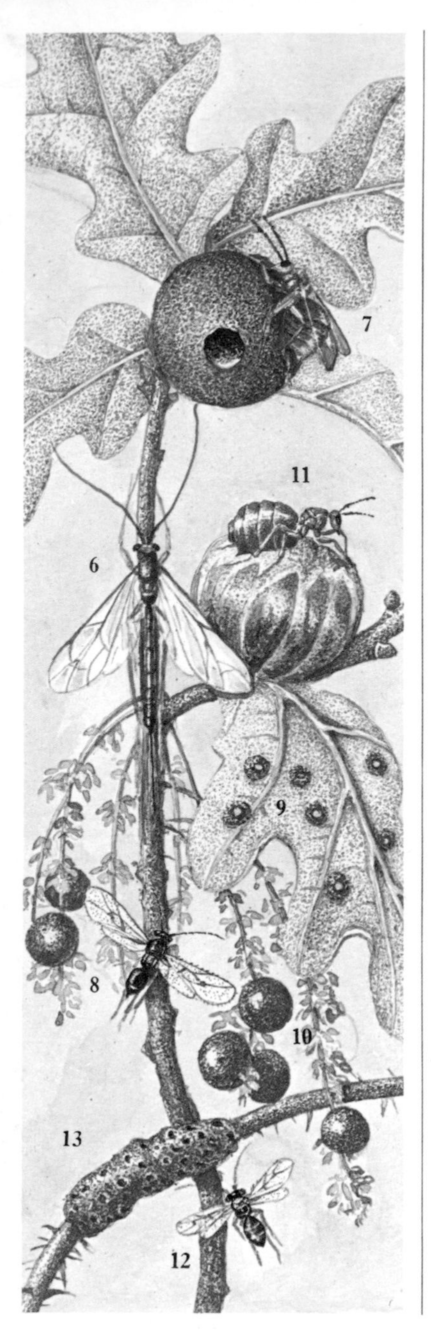

Jet Ant *Lasius fuliginosus* FORMICIDAE BL 3–5 mm. Shiny jet-black, with yellowish tips to legs. Gives off vinegary smell if touched. Diurnal, Jul–Aug; common in woods, pastures. Predatory; takes honeydew from aphids. Colonies nest in tree-trunks or make small mounds on ground. T, ex Ic. [♀ **1**]

Yellow Ant *Lasius flavus* FORMICIDAE BL 3–4 mm. Pale yellow or brownish. Bites painfully if touched. Summer; widespread in woods. Colonies live underground; often form mounds <2 m long. Keeps aphids on grass roots, 'milking' them for honeydew. T, ex Ic. [☿ **2**]

Bee-killer Wasp *Philanthus triangulum* SPHECIDAE BL 12–15 mm. Head broad. Diurnal, Jul–Aug; local in sandy areas, downs. Attacks honey-bees, often waiting for them on fls; paralyses them with sting, stores in nest as food for larvae. WE, CE. [**3**]

Field Digger-wasp *Mellinus arvensis* SPHECIDAE BL 9–15 mm. Abdomen enlarged towards rear. Diurnal, Jul–Sep; widespread on sandy soils in agricultural areas, grasslands. Catches flies by leaping on and stinging them. Nest tunnel in sandy ground leads to brood cells stocked with paralysed flies on which larvae feed. T, ex Ic. [**4**]

Hunger Wasp *Gasteruption assectator* GASTERUPTIONIDAE BL 8–12 mm. Slender with sickle-shaped abdomen broader towards rear, swollen segment on hindlegs; black with 3 (♀) or 4 (♂) orange marks on each side of abdomen. Diurnal, Jun–Aug; local in grasslands, hedgerows. Often on fls, esp umbellifers. Parasitizes other bee spp. T, ex Ic. [♀ **5**]

ichneumon fly, parasitic wasp *Netelia testacea* ICHNEUMONIDAE BL 16–17 mm, ♀ with long ovipositor. Legs long; thorax and abdomen slender, orange-brown with darker tip to abdomen. Does not sting man. Diurnal, also attracted to light, summer; widespread in woods, hedgerows.

Eggs laid in caterpillar, esp noctuid moth; larva develops inside caterpillar, eventually killing it. T, ex Ic. [**6**] Many similar spp.

Marble Gall-wasp *Andricus kollari* CYNIPIDAE BL 4–5 mm. Yellow-brown, with clear wings. Diurnal, May–Jun, Aug–Sep; common in woodlands. Eggs laid in oak buds in summer cause formation of hard, round marble-galls ('oak-apples'), <18 mm across, from which only ♀s emerge, Aug–Sep; these lay eggs in buds of oaks, esp Turkey oaks, resulting in soft, oval galls, <3 mm long, obvious in Apr, from which both ♂s and ♀s emerge, May–Jun, to mate and restart cycle. WE, CE. [♀ on gall **7**]

Common Spangle or **Oak-spangle Gall-wasp** *Neuroterus quercus-baccarum* CYNIPIDAE BL 4–5 mm. Diurnal, Mar–Apr, Jun–Jul, common in woods. Eggs laid on oak lvs in summer cause circular, lens-shaped galls on underside – yellow-green when young, red later with growth of red hairs – each containing 1 larva; galls overwinter in leaf-litter, ♀s emerge Mar–Apr to lay eggs on young oak lvs or catkins; resulting galls show in May–Jun as red, berry-like growths ('currant-galls'), from which both ♂s and ♀s later emerge. T, ex Ic. [**8**] [spangle-galls **9**] [currant-galls **10**]

Oak-apple Gall-wasp *Biorrhiza pallida* CYNIPIDAE BL *c*5 mm. ♂ yellow-brown, with reddish base to abdomen; ♀ wingless, with black lines across abdomen. Diurnal, Jun–Jan; common in woods, hedgerows. Eggs laid, Dec–Jan, in oak buds which then turn yellowish and swell into spongy, reddish-tinged (later brown) 'oak-apples' <4 cm across; each has several chambers holding larvae. T, ex Ic. [♀ on gall **11**]

Bramble Gall-wasp *Diastrophus rubi* CYNIPIDAE BL 2–3 mm. Shiny black, with brown legs, transparent wings. Diurnal, May–Jun; common in woods, hedgerows. ♀ lays eggs under surface of bramble or raspberry stem. Larva overwinters in stem, causing spindle-shaped lumps on surface. T, ex Ic. [**12**] [gall **13**]

Beetles (Coleoptera). Very small to large, mostly compact, with tough horny covering, biting mouthparts, usually strong legs, no cerci (*cf* cockroaches); forewings form non-overlapping, horny elytra which protect folded hindwings and abdomen. Many fly readily; others flightless. Metamorphosis complete. Larvae varied in shape, some legless.

ground-beetle *Harpalus rufipes* CARABIDAE BL 14–16 mm. Thorax narrowed at rear, with finely speckled surface there; elytra closely spotted between grooves. Rapid runner. Mainly nocturnal, spring-summer; common in hedgerows, woods, mostly in lowlands. Feeds on invertebrates, also partly vegetarian. Larva elongate, predatory; under leaf-litter. T, ex Ic. [**1**]

ground-beetle *Zabrus tenebrioides* CARABIDAE BL 14–16 mm. Rather hump-backed shape, thorax not obviously narrowed. Crepuscular, spring-summer; local in agricultural land. Climbs corn stalks to eat grain; can be destructive. Larva feeds nocturnally on young corn lvs. T, ex Ic, Fi, No, rare Sw. [**2**]

ground-beetle *Carabus nemoralis* CARABIDAE BL 20–24 mm. Head and thorax rather narrow; thorax and elytra sheened. Flightless. Mostly nocturnal, Apr–Sep; widespread in grasslands, downs, meadows. Preys on other insects. Larva predatory; lives under stones. T, ex Ic. [**3**]

carrion-beetle *Oiceoptoma thoracicum* SILPHIDAE BL 13–16 mm. Rather flattened, strongly sculptured. Spring-summer; widespread in grasslands, downs, hedgerows. Feeds on carrion, also fungi, rotting wood. Larva squat, like woodlouse but with fewer legs; also eats decaying animal and plant material. T, ex Ic. [**4**]

carrion-beetle *Dendroxena quadrimaculata* SILPHIDAE BL 10–14 mm. Head small. Nocturnal, Apr–Jul; local in woods, hedgerows, esp on oaks, beech, fruit trees. Does not eat carrion like rest of family, but preys on caterpillars. Larva elongate; also eats caterpillars. T, ex Ic, rare FS. [**5**]

burying or **sexton-beetle** *Necrophorus vespilloides* SILPHIDAE BL 10–15 mm. Antennae entirely black (club reddish-yellow in rest of genus). Spring–summer; widespread in agricultural areas, grasslands. Adult attracted to carrion; excavates ground under dead animal, which slowly sinks till buried; eggs laid in corpse; larva hatches and feeds below ground. T, ex Ic. [**6**] Several similar spp.

rove-beetle *Anotylus tetracarinatus* STAPHYLINIDAE BL 1·5–2 mm. Elytra short. Spring–summer, sometimes all year; common, often abundant, in pastures, agricultural land, living in haystacks, dung, any refuse. Larva in similar places. Adult flies freely; 'fly in eye', if not thrip, is usually this beetle. T, ex Ic. [**7**] Many similar spp.

rove-beetle *Philonthus laminatus* STAPHYLINIDAE BL 8–11 mm. Body narrow and elytra short like other rove-beetles, distinguished by combination of colours and head narrower than thorax. Spring–summer; widespread in grasslands, woods, hedgerows. Adult and larva in dung or decaying vegetation. T, ex Ic. [**8**]

Devil's Coach-horse, Cock-tail *Staphylinus olens* STAPHYLINIDAE BL 20–28 mm. Short, stumpy elytra. Nocturnal, spring–summer; common in grasslands, downs, agricultural areas, hiding by day under stones. If disturbed, erects hind end and opens jaws in threatening attitude, emitting foetid smell. Predator; can give sharp bite. Larva in decaying vegetation. WE, CE, not Ic, Fi. [**9**]

Glow-worm *Lampyris noctiluca* LAMPYRIDAE BL ♂ 10–13 mm, ♀ 10–18 mm. Front of thorax semicircular, covering head, with transparent spot over eyes; ♂ elytra gape apart, exposing apex of abdomen; ♀ wingless, undersides of last 3 abdominal segments strongly luminescent. ♀ glows to attract ♂ (egg, pupa and ♂ also glow slightly). Nocturnal, Jun–Jul; local in grasslands, hedgerows. Larva sucks liquid from paralysed slugs, snails. T, ex Ic. [♀ **10**]

5
2
9
10
3
1
6
4
7
8

soldier-beetle *Rhagonycha fulva* CANTHARIDAE BL 10–11 mm. Yellow-red, with black tips to elytra. Very active, flies freely in sunshine, Aug–Sep; widespread, common in woods, often on umbelliferous fls with other spp of soldier-beetles. Carnivorous. Larva predatory, often in rotting wood. T, ex Ic, local Fi. [1]

click-beetle *Ampedus sanguineus* ELATERIDAE BL 10–15 mm. Elytra scarlet, ribbed. If falls on back, bends body backwards and springs up with audible click. Aug–Oct; common in old tree stumps in woods, also frequenting umbelliferous fls. Larva or 'wireworm' elongate, slender; feeds on rotting wood, plant material. T, ex Ic. [2]

click-beetle *Ctenicera pectinicornis* ELATERIDAE BL 13–17 mm. Antennae of ♂ very feathery, of ♀ less so. Clicks like *A. sanguineus* when righting itself. Aug–Sep; widespread in wood edges, hedgerows. Larva or 'wireworm' associated with rotting wood. T, ex Ic. [♂ **3**] Related spp important agricultural and horticultural pests.

beetle *Dicerca berolinensis* BUPRESTIDAE BL 20–23 mm. Elytra pointed at rear. Active in sunshine, summer; woodlands, esp beechwoods, on decaying beech, alder, hornbeam. Larva feeds in decaying wood. Fr, Lu, Be, Ge, Cz, Po. [**4**]

Two-spot Ladybird *Adalia bipunctata* COCCINELLIDAE BL 3–5 mm. Elytra very variable, from red with 2 or more black spots or black pattern, to largely black with red patches. Spring–summer; hibernating through winter; widespread, common in hedgerows, woods, grasslands. Preys on aphids. Larva elongate, also feeds on aphids. T. [**5**] Several other common spp, some yellow-and-black, with different numbers of spots.

beetle *Oedemera nobilis* OEDEMERIDAE BL 8–10 mm. Slender, green with whitish-yellow lines; ♂ thorax striped, hind femora enlarged; ♀ thorax with light patch in centre, femora normal. Jun–Jul; widespread, common on fl-heads in hedgerows, grasslands. Larva elongate; feeds in plant stems, possibly also predatory. Br, Ir, Fr, Lu, Be, Ne, Ge, Cz, Po, local De. [♂ 6]

cardinal-beetle *Pyrochroa coccinea* PYROCHROIDAE BL 13–15 mm. Antennae with toothed edge; thorax and elytra scarlet. Jun–Jul; widespread in woods, hedgerows, usually living under dead bark but also on fls, esp at wood edges, in warm weather. Larva feeds in rotting wood for several years. T, ex Ic. [7] Several similar spp.

Bluish Oil-beetle *Meloe violaceus* MELOIDAE BL 14–36 mm, ♂ much smaller than ♀ swollen with eggs. Head sharply constricted behind eyes; thorax sculptured with fine marks; elytra very short, exposing abdomen, and slightly overlapping; all violet-blue with metallic sheen. Flightless, slow-moving; exudes evil-smelling fluid from joints if disturbed. Summer; widespread, common in grasslands, downs. ♀ lays thousands of eggs in cracks in ground; larva climbs to flower, tries to attach itself to visiting bee; if carried into nest, feeds on eggs, larvae, honey. T, ex Ic. [8] *M. proscarabaeus* very similar in appearance and life-history, but more metallic purple.

beetle *Anaspis maculata* SCRAPTIIDAE BL 2–3 mm. All yellow-brown, with conspicuous black marks on elytra. Flies readily. Spring–summer; common in hedgerows, grasslands, often on fls. Larva scavenges in rotting wood. Br, Fr, Lu, Be, Ne, De, Ge, Cz, Po. [9]

dor-beetle *Geotrupes spiniger* SCARABAEIDAE BL 16–25 mm. Compact, with powerful legs; all black. Slow-moving. Flies early evening, spring–summer, some overwinter; widespread in grasslands, cultivation, associated with dung or rotting fungi. Larva in dung. T, ex Ic, Fi, No. [10] *G. stercorarius* similar in appearance and habits, but shinier, with spotted, hairy abdomen.

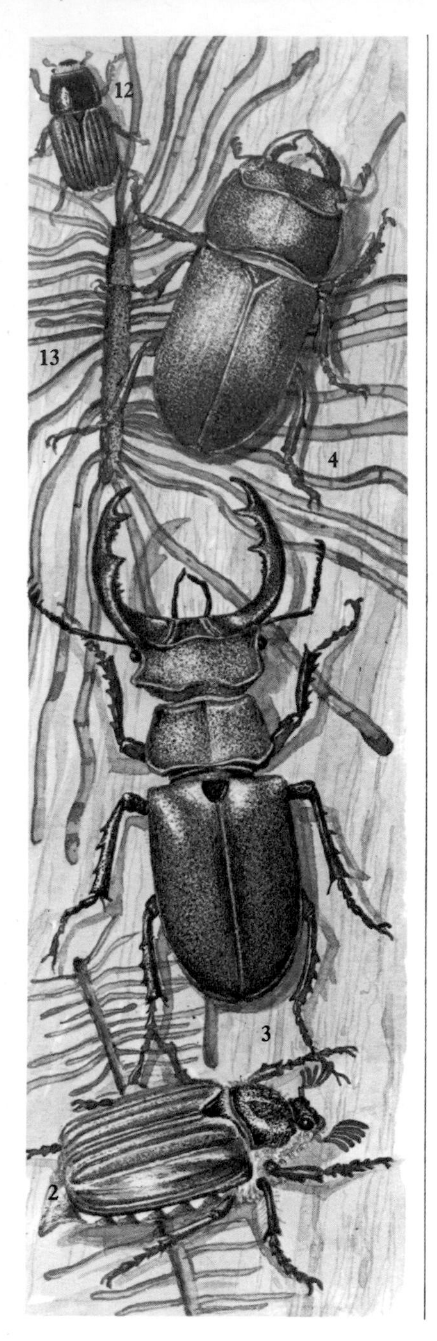

dung-beetle *Aphodius rufipes* SCARABAEIDAE BL 10–12 mm. Summer; widespread in cultivated areas, grasslands, downs, associated with dung or rotting vegetation. Larva in dung. T, ex Ic. [**1**]

Cockchafer, Maybug *Melolontha hippocastani* SCARABAEIDAE BL 23–25 mm. Antennae of ♂ distinctive, of ♀ much smaller. Hums in flight. Crepuscular, May–Jun; common in grasslands, cultivation. Feeds on lvs of various trees. Larva white grub; feeds on roots of many herbs, shrubs, trees, very destructive to grasses, crops. T, ex Ic, but local Br, Ir. [♂ **2**] *M. melolontha*, similar, but tip of abdomen longer, narrower, more pointed.

Stag Beetle *Lucanus cervus* LUCANIDAE BL 20–50 mm, ♂ much larger than ♀. ♂ only has huge, antler-like (but weak) jaws protruding well beyond head; 3 teeth on hind tibiae. Crepuscular, attracted in flight to lights, summer; local in woods, wood edges, parklands. Larva feeds in rotting wood. WE, CE, not Ic, Fi. [♂ **3**]

Lesser Stag Beetle *Dorcus parallelipipedus* LUCANIDAE BL 18–30 mm. Mandibles of ♂ prominent and curved, of ♀ less strongly developed and straight; 1 tooth on hind tibiae. Nocturnal, summer; widespread in woods, wood edges, esp on ash. Larva in rotten wood. T, ex Ic, Fi, Sw, ?No. [♂ **4**]

Musk Beetle *Aromia moschata* CERAMBYCIDAE BL 20–30 mm. Antennae long, thorax pointed at sides; red-brown spots at bases of elytra; sometimes coppery, even bright blue. Emits smell recalling musk. Jul–Aug; local in woodlands, associated with willows. Larva in rotting wood. T, ex Ic. [**5**]

leaf-beetle *Oulema melanopa* CHRYSOMELIDAE BL 4–5 mm. Summer–autumn; widespread in agricultural areas, grasslands. Feeds on plants, sometimes pest of cereals. Larva also plant-feeder. T, ex Ic. [**6**] Many other spp of leaf-beetles, some colourful, often pests.

leaf-beetle *Chrysolina staphlea* CHRYSOMELIDAE BL 6–9 mm. Compact, rounded. Summer; widespread, common in grass, low vegetation, in pastures, downs, cultivation. Adult and larva both plant-feeders. T, ex Ic. [7] Many allied spp.

Colorado Beetle *Leptinotarsa decemlineata* CHRYSOMELIDAE BL 11–12 mm. Late spring–Oct, some overwintering; in agricultural areas, where both adult and larva serious potato pests capable of stripping crop of lvs. Hibernates in soil, at depth <30 cm; on emerging, flies to potato field, often several km away, feeds on lvs, ♀ lays clusters of yellow eggs on undersides. Presence in Br must be notified to authorities. (WE, CE, sSw, rare Fi, accidental Br, Ir, often in grain shipments; from N America.) [8] [larva 9]

Pea Weevil *Sitona lineatus* CURCULIONIDAE BL 4–5 mm. Antennae slightly elbowed, with small apical club; rostrum square-ended, furrowed at top. Spring–autumn, also hibernating in stubble, farm-buildings; lays eggs in soil round base of pea plant. Larva burrows into root nodules; pupates after 6–7 weeks. Emerging adult feeds on pea or bean plant, causing U-shaped notches in lvs; later migrates to clover or lucerne. T, ex Ic. [10]

weevil *Apion miniatum* CURCULIONIDAE BL 3·5–4·5 mm. Summer; common in grasslands, agricultural areas. Adult and larva feed on docks. T, ex Ic. [11] Many spp of weevils, mostly characterized by elongated rostrum.

Elm-bark Beetle *Scolytus multistriatus* SCOLYTIDAE BL 3–6 mm. Apr–May, Jul–Aug; widespread on elms. ♀ burrows under bark of elm to lay eggs in central tunnel, larvae excavate 'fish skeleton' side-galleries. Damages trees and, with similar large elm-bark beetle *S. scolytus*, responsible for transmission of fungus causing Dutch elm disease. T, ex Ic. [12] [galleries 13] Other spp bore under bark of various other trees.

Arachnids (spiders, harvestmen, ticks etc.) all have 4 pairs of legs. Spiders (Araneae), have 2 divisions to the body, cephalothorax (fused head and thorax) and abdomen, joined by a narrow waist; cephalothorax covered by shield known as carapace; silk glands opening from spinnerets on abdomen. Prey on small invertebrates; *c*½ spp use silk for building webs to trap pray; others hunt.

purse-web spider *Atypus affinis* ATYPIDAE BL 8–12 mm. ♀ robust with heavy jaws, short-legged; ♂ less robust, dark brown. Adult ♀ all year, ♂ in spring; chalk grassland, meadows, sunny places. ♀ lives in partly buried, camouflaged, sealed tube; insects walking over tube are impaled by fangs and pulled in, and tear is then repaired. T, ex Ic, Fi, Sw, No. [♀ **1**]

crab-spider *Xysticus cristatus* THOMISIDAE BL 4–7 mm. Carapace dark brown with median light yellow band; abdomen light yellow with brownish, leaf-like markings, darker on ♂. Adults May–Jul; common in grassland, woodland. Builds no snare, but lurks in low vegetation, grabs unwary insects. T. [♀ **2**]

spider *Evarcha falcata* SALTICIDAE BL 5–7 mm. ♂ black with white markings, central reddish spots and stripes on abdomen. Adults May–Jul; frequent on low vegetation, in open woodlands, jumps from plant to plant. Spins no snare, but hunts prey visually. T, ex Ic. [♀ **3**]

spider *Pisaura mirabilis* PISAURIDAE BL 12–15 mm. Variable colour, large size, tapered abdomen and resting posture distinctive. Adults Apr–May; common in long grass. Hunts prey, carries egg-cocoon in jaws, builds tent-like covering for young to hatch. T, ex Ic. [**4**]

pirate-spider *Ero furcata* MIMETIDAE BL 2·5–3·5 mm. Carapace yellow-brown with darker markings; abdomen ornamented with black, red, yellow and white. Adults Apr–Oct; grassland, low plants and bushes. Moves very slowly, feeds on other spiders, invading webs. Egg-cocoons distinctive, globular, surrounded with tangled brown threads and suspended from vegetation. T, ex Ic. [**5**] 2 other very similar spp.

spider *Theridion pallens* THERIDIIDAE BL 1–2 mm. Adult ♂ Apr–May, ♀ Apr–Sep; frequently in fine webs of tangled threads on undersides of lvs, esp oaks and evergreens. ♀ often with peculiar egg-cocoon. T, ex Ic, No. [**6**]

spider *Tetragnatha montana* TETRAGNATHIDAE BL 6–11 mm. Slender, with long, tapered abdomen, long legs and protruding jaws. Adults May–Jun; widespread on shrubs, trees, tall grasses, often near water. Spins fine orb-web; at rest lies along stems with outstretched legs. T, ex Ic, Fi. [**7**] Several similar spp.

spider *Meta segmentata* TETRAGNATHIDAE BL 5–8 mm. Variable; carapace yellow-brown with dark, wedge-shaped mark; abdomen whitish-grey with characteristic, toothed band, wider in front with grey, black, yellow and reddish markings. Adults Aug–Sep; abundant on low vegetation in woodland, hedgerows. Spins orb-web. T, ex Ic. [♀ **8**] *Meta mengei* very similar, but adults Apr–Jun. T, ex Ic.

spider *Araneus umbraticus* ARANEIDAE BL 11–14 mm. Distinctive, large, flat spider adapted to living in crevices. Adults Jun–Oct; under loose bark of trees and posts; nocturnal. Spins large orb-web, but spider difficult to find. T, ex Ic. [♀ **9**]

hammock-spider *Linyphia triangularis* LINYPHIIDAE BL 5–6 mm. Carapace with distinctive, median, bifurcated line; legs with greenish tinge; ♀ abdomen with median, light to dark brown, dentate band, bordered by light yellow. Adults often together Sep–Oct; common in shrubs, hedges, tall grasses, in gardens, hedgerows, woodland. Builds no retreat, lives inverted on underside of hammock-web. T, ex Ic. [♂ **10**]

2
6
5
9
7
8
4
3
10
1

Amphibians are cold-blooded; in winter dormant on land or at bottom of pond. Life cycle with metamorphosis; larvae aquatic: miniature adults with gills (newts, salamanders) or tadpoles (frogs, toads). Adults in damp, shady places, most active at night; during breeding season newts, frogs, toads in ponds when most often seen. Frogs, toads: breeding ♂s call in chorus, ♂ clasps ♀ in water, while ♀ lays eggs in gelatinous clumps or strings ♂ discharges sperm over them (external fertilization). Newts also mate in water: elaborate courtship display, then ♂ drops sperm in package which ♀ picks up and places inside her (internal fertilization). Salamanders have internal fertilization, but mate on dry land: ♂ grasps ♀ from below, deposits sperm package so that she falls on to it; eggs develop within ♀, larvae born into water. Long-lived, <20 years. Endangered by pollution and filling in of ponds.

Common Frog *Rana temporaria* RANIDAE
L 7–9 cm. Smooth skin; colour variable, pale with dark markings; dark pads on fore-limbs of breeding ♂. Soft, purring croak during mating. Damp places in fields, meadows, woods, gardens. Breeds Mar–Apr; migrates to ponds, often forming large aggregations; eggs laid in clumps. Food insects, worms. T, ex Ic. [1]

Springing Frog *Rana dalmatina* RANIDAE
L 5–7 cm. Pointed snout, very long hind-limbs; smooth skin; pale brown with dark markings, inverted V behind head. Weak, chirp-like call. Very powerful jumper. Damp places in open deciduous woodlands, banks, meadows. Breeding follows spring thaw; eggs laid in clumps which float. Food insects, worms. Fr, De, Ge, Cz, sSw. [2]

Common Toad *Bufo bufo* BUFONIDAE
L 8–12 cm. Broad, plump; short, rounded head; warty skin; colour variable, brown, green, reddish. Deep, croaking call during mating. Damp places in woods, gardens, fields. Breeds Mar–May; migrates to ponds, often forming large aggregations;

eggs laid in strings, twined around weeds. Food insects, worms, slugs. T, ex Ir, Ic, nFS. [3]

Green Toad *Bufo viridis* BUFONIDAE L 6–9 cm. Broad, plump; short, rounded head; warty skin; grey or pale green with dark green markings and red warts; becomes darker in the shade. Long, trilling whistle. Coastal and sandy areas, dikes, ditches; hides by day under stones. Breeds Apr–Jun; eggs laid in long strands. Food insects, worms. De, Ge, Cz, Po, sSw. [4]

European Salamander *Salamandra salamandra* SALAMANDRIDAE L 18–24 cm. Stout body, plump legs, vertical grooves on body, large glands on head; skin glossy, black with yellow or orange spots or stripes. Shady, damp places in hilly country with deciduous woods. Mates during spring and summer; ♀ goes to pond to deliver larvae. Food insects, worms, slugs. Skin produces a poisonous secretion. T, ex Br, Ir, Ic, De, FS. [5]

Smooth Newt *Triturus vulgaris* SALAMANDRIDAE L 6–10 cm. Colour variable; ♂ in breeding season has denticulated dorsal crest and tail, body pale green, brown or grey with black spots, orange belly, blue and red stripes at base of tail; ♀ in breeding season pale yellow or brown, yellow belly, small dark spots all over. In spring found in ponds, ditches; rest of year on land, in damp places, *eg* under logs. Mates Feb–Jun; ♀ wraps eggs individually in leaves. Food small insects, frog spawn, tadpoles. T, ex Ic, nFS. [♂ 6] [♀ 7]

Carpathian Newt *Triturus montandoni* SALAMANDRIDAE L 7–10 cm. Colour variable; ♂ in breeding season has low, dorsal crest, filament at end of tail, body yellow, green or brown above, marbled or spotted darker, belly yellow to red, row of dark spots along lower edge of tail; ♀ similar to ♀ smooth newt. Found in damp places in hilly, wooded country <1000 m. Mates Mar–Jun. Food small insects, other invertebrates. Cz. [♂ 8]

Reptiles are cold-blooded; but can control body temperature to a limited extent, physiologically and behaviourally, *eg* by basking in sun. (Snakes often in warm places, *eg* under iron sheets.) Need to raise body temperature in morning before able to resume activity. Lizards with legs, moveable eyelids; will drop tails if grabbed (autotomy), new tails unlike originals in structure or colour. Snakes with fixed eyelids, loose jaw hinges for swallowing prey whole; moult regularly, discarding skin intact. Of the N European snakes, only adder (viper) is poisonous. Internal fertilization: ♂ with penis. Eggs with shells, laid in dry places, fewer than in amphibians; or young born alive. Young are like miniature adults. Endangered because of clearance of open heathland habitats.

Green Lizard *Lacerta viridis* LACERTIDAE L 30–40 cm. ♂, back shiny green spotted black, belly yellowish, throat blue; ♀ brown with longitudinal pale stripes. Agile climber. Dry places on scrub, heathland, around coniferous woodland. Hibernates Oct–Apr. Breeds May–Jun. Food chiefly insects and larvae, also fruit. Fr, Ge, Cz, Po. [♂ 1]

Wall Lizard *Podarcis muralis* LACERTIDAE L 18–20 cm. Pointed nose; markings variable, usually a number of longitudinal stripes. Agile climber. Dry, sunny places; rocky cliffs, garden walls. Hibernates Oct–Apr. Breeds May–Jun. Food insects. Fr, Lu, Be, Ne, Ge, Cz. [2]

Grass Snake *Natrix natrix* COLUBRIDAE L 70–150 cm. Colour variable, upperside dark brown or green, with yellow or white 'collar' and jaws. Good swimmer. Dense undergrowth near water in open woodlands, farmland, hedges. Hibernates Oct–Apr. Mates Apr–May, eggs laid Jul–Aug. Food frogs, toads, newts. When grasped secretes a noxious slime. T, ex nBr, Ir, Ic, FS n of 67°N. [3]

European Whip Snake *Coluber viridiflavus* COLUBRIDAE L 150–190 cm. Slender body; upperside black with yellow-green markings, latter forming transverse bands towards anterior end, longitudinal stripes posteriorly; underside yellowish-grey. Fast moving, good climber. Sunny areas with bushy vegetation, stone walls, vineyards. Hibernates Oct–Apr. Mates Apr–May, eggs laid Jul. Food birds' eggs, lizards, frogs, small rodents, insects. Very aggressive. Fr. [4]

Aesculapian Snake *Elaphe longissima* COLUBRIDAE L 100–200 cm. Slightly flattened body; colour variable, usually olive-brown above, yellowish below. Agile climber. Among stones in rocky areas in open deciduous woodland. Hibernates Oct–May. Mates May, eggs laid Jul. Food mice, birds' eggs, nestlings. Fr, Ge, Cz, Po. [5]

2
1
3
4
5

Black Stork *Ciconia nigra* CICONIIDAE L 90–105 cm. Resembles white stork, but glossy black with white underbody; imm browner, esp head and neck, with grey-green bill and legs. More vocal than white stork when nesting, but bill-clatters less. Undisturbed forests interspersed with streams, pools, wet meadows. Nest of sticks, lined moss, in tall tree, 3–5 eggs, Apr–Aug. Food fish, frogs, also insects, small reptiles and mammals, leeches. Summer Cz, Po, rare Ge. [ad **1**] [imm **2**]

White Stork *Ciconia ciconia* CICONIIDAE L 95–110 cm. Large, with long neck extended in flight, short tail; all-white but for black flight-feathers, red bill and legs. Loud bill-clattering at nest. Grasslands, arable fields, wetlands, often breeding on houses, even in towns. Sometimes colonial; bulky nest of sticks, in tree, on building, telegraph pole, 3–5 eggs, Apr–Aug. Food frogs, large insects, voles, shrews, lizards. Summer De, Ge, Cz, Po, now rare Fr, Ne; more widespread on passage. [3]

Honey Buzzard *Pernis apivorus* ACCIPITRIDAE L 50–60 cm. Head small, pigeon-like, protruding in flight; wings broad, narrow-based; tail longish, dark band at end, 2 narrow bands near base. Whistling and grating notes, esp at nest. Breeds in open woods, esp broadleaved. Smallish nest of sticks, lined green sprays, high in beech or other tree, 1–3 eggs, May–Sep. Food wasp grubs, bees and honey, ant pupae, also other insects, worms, frogs, reptiles, mice, birds' eggs, berries. Summer sBr, Fr, Lu, Be, Ne, De, Ge, Cz, Po, sFS. [4]

Black Kite *Milvus migrans* ACCIPITRIDAE L 55–60 cm. More compact than red kite, with tail shorter, less forked, triangular when spread. Thin, whinnying, gull-like squeal, chattering notes, esp when nesting. Cultivated areas with trees, open woodland, esp near lakes, rivers. Often social; small to large nest of sticks, dung, moss, paper, rags, in tree, 2–3 eggs, May–Aug. Food dead fish, other carrion, garbage, small mammals, reptiles, frogs, pond snails, insects. Summer Fr, Lu, Ge, Cz, Po, sFi; passage Be, Ne, De, sSw. [5]

Red Kite *Milvus milvus* ACCIPITRIDAE L 60–65 cm. Wings long, fairly broad, set well forward on slim body with long, forked tail; underwings with white patch at base of primaries. Buzzard-like mewing, also whinnying recalling black kite. Open, mature woodland, or parkland and cultivation with scattered trees, also wooded hills. Flat nest of sticks, earth, rags, paper, in broadleaved tree, 2–3 eggs, Mar–Jul. Food small mammals, young birds, reptiles, frogs, stranded fish, carrion, insects. Resident wBr: Wales; summer Fr, Lu, Ge, Cz, Po, sSw, irregular Be, De; passage Be, Ne, De. [**6**]

Buzzard *Buteo buteo* ACCIPITRIDAE L 50–56 cm. Compact, head short, wings broad, tail short; varies from largely whitish to black-brown beneath. Characteristic, plaintive, ringing mew. Wooded lowlands to moorland crags, sea-cliffs. Flat nest of sticks, bracken, grass, decorated with green sprays, in tree or on cliff, 2–3 eggs, Mar–Jul. Food small mammals, carrion, earthworms, also birds, reptiles, amphibians, insects. Summer T, ex Ic, nFS, but only extreme nIr; winter T, ex Ic, Fi, No, but only extreme sSw. [**7**]

Rough-legged Buzzard *Buteo lagopus* ACCIPITRIDAE L 50–60 cm. Distinguished from buzzard by longer, black-ended, white tail and elastic beats of longer wings; also by dark patch on belly. Call lower-pitched, more miaowing. Tundra, moorland, northern forests, but winters open lowlands, downs, heaths, marshes, sand-dunes. Food mammals to size of rabbit, also birds. Summer nFS; scarce winter eBr, Lu, Be, Ne, De, Ge, Cz, Po, sSC; passage Fr, sFi. [**8**]

Booted Eagle *Hieraaetus pennatus* ACCIPITRIDAE L 45–55 cm. Buzzard size, but head and tail longer, body and wings narrower; light and dark colour phases. High, whistling 'ki-keee'. Broadleaved woods, mainly on hillsides, hunting over open country. Large nest of sticks, lined green lvs, high in tree, sometimes on rocky ledge, 1–2 eggs, Apr–Aug. Food mainly small mammals, also birds. Summer Fr, Cz, ?ePo. [light **9**] [dark **10**]

1
2
3
4
5
6
7
8
9
10

Short-toed Eagle *Circaetus gallicus* ACCIPITRIDAE L 63–70 cm. Large, head broad and protruding, wings long and broad with spread primaries; below, white wings with blackish lines, 3 bands on tail, usually dark head. Mostly silent, but musical whistling, buzzard-like mewing, when nesting. Lightly wooded lowlands, grassy plains, mountain slopes. Nest of thin twigs, lined green sprays, near top of low tree, rarely on cliff ledge, 1 egg, May–Aug. Food snakes, also lizards, small mammals. Summer Fr, eCz, ePo. [1]

Lesser Spotted Eagle *Aquila pomarina* ACCIPITRIDAE L 60–68 cm. Medium-sized in flight, without particularly protruding head, broad wings or short tail; mainly dark brown, but for grey-brown head and mantle, whitish areas on tail-coverts above and below; imm more uniform brown, but bold white U on uppertail-coverts, wings with whitish line along middle, some spotting on coverts. Call higher pitched, more yelping than spotted eagle. Lowland woods by open country, not necessarily near water. Nest of sticks, lined grasses, green sprays, in tall tree, 1–3 eggs, Apr–Aug. Food small mammals, lizards, frogs, birds, insects. Summer Ge, Cz, Po. [2]

Spotted Eagle *Aquila clanga* ACCIPITRIDAE L 65–75 cm. Bulky; in flight distinguished from lesser spotted by broader wings bulging at rear, shorter tail, smaller head, darker; imm almost black, with similar white markings, but also rows of large spots on wing-coverts. Dog-like yapping, mainly when breeding. Wooded river valleys, marshes with woods or scattered trees. Nest like lesser spotted, 1–2 eggs, Apr–Aug. Food frogs, reptiles, fish, also small mammals, birds, carrion. Summer eCz, ePo, swFi, but scarce; passage Ge, wCz, wPo, sFi, sSw. [3]

Lesser Kestrel *Falco naumanni* FALCONIDAE L 29–32 cm. Very like kestrel, but slimmer, with narrower wings, thinner wedge-ended tail; ♂ head and tail bluer, no moustache, upperparts brighter chestnut without spots, blue area also on wings, underbody less spotted, underwings cleaner white with blacker

tips, whitish claws. Hovers much less. Chattering 'kik-kik-kik' when breeding. Open country, chiefly lowlands, also towns, breeding in cliffs, ruins, large buildings. Colonial; nest scrape in hole, 4–5 eggs, May–Jul. Food mainly large insects, also lizards, some small mammals, frogs. Summer eCz, sePo. [♂ 4] [♀ 5]

Kestrel *Falco tinnunculus* FALCONIDAE L 33–36 cm. Small, wings pointed, tail long, hovers frequently; ♂ head grey with black moustache, upperparts chestnut spotted with black, tail grey with black band. Shrill, chattering 'kee-kee-kee' near nest. Almost ubiquitous, even in towns. Nest scrape in old crow nest, hole in tree, cliff, building, 4–5 eggs, Mar–Jul. Food mice, voles, small birds, also insects, earthworms, lizards, frogs. T, ex Ic, but only summer ePo, Fi, nSC. [♂ 6] [♀ 7]

Red-footed Falcon *Falco vespertinus* FALCONIDAE L 28–32 cm. Small, wings pointed; ♂ sooty-grey but for dull red thighs and undertail, orange-red bill and legs; ♀ reddish-sandy crown and underbody, barred grey upperparts and tail; juv closely barred above, heavily streaked below, with pale forehead. Call kestrel-like, but shriller, more drawn-out. Grasslands, cultivation, with scattered clumps of trees. Colonial in rookeries; nest scrape in old nest of rook, magpie or crow, in tree, 3–5 eggs, May–Aug. Food mainly insects; when nesting, also lizards, frogs, small mammals. Summer eCz, sePo, but very scarce; more widespread on passage, to Ge, wPo. [♂ 8] [♀ 9]

Hobby *Falco subbuteo* FALCONIDAE L 30–36 cm, ♂ smaller than ♀. Medium-sized, tail short, but wings long, narrow, scythe-like, sharply pointed, recalling huge swift; note reddish 'trousers'; juv darker brownish-black above, thighs and undertail less red, more buff. Noisy near nest, esp with young, including loud, clear 'kew-kew-kew'. Open country with trees. Nest scrape in old crow or raptor nest in tree, esp pine, 2–3 eggs, Jun–Aug. Food small birds, large insects, also bats. Summer sBr, Fr, Lu, Be, Ne, De, Ge, Cz, Po, sFi, sSw. [10]

Red-legged Partridge *Alectoris rufa* PHASIANIDAE L 33–36 cm. Distinguished from smaller grey partridge by white supercilia, black-bordered white throat, lavender flanks barred chestnut, black and white, red bill and legs. ♂ harsh 'chucka-chucka'. Fields, pastures, downs, scrubby wastes, also sand-dunes, coastal shingle. Nest hollow lined dead grasses, lvs, in hedgerow, crop, long grass, 10–16 eggs, Apr–Aug. Food mainly grass lvs, clovers, cereals, weed seeds, roots. wFr, (sBr). [**1**]

Grey Partridge *Perdix perdix* PHASIANIDAE L 29–32 cm. Rounded, chicken-like, wings and tail short; whole face orange-brown, foreneck grey, flanks chestnut bars broken with whitish cross-streaks; ♂ has inverted, dark chestnut horseshoe on lower breast, much less developed on ♀. Whirring flight, gliding on bowed wings. ♂ grating 'kirr-ic'. Both arable and grass, with hedges, scrub and waste ground, also downs, heaths, moorland edges, sand-dunes. Nest hollow lined grasses, lvs, at base of hedge, also in nettles, mowing grass, clover, corn, 9–20 eggs, Apr–Aug. Food grass lvs, clovers, weed seeds, grain, buds, roots, insects in summer, also earthworms, slugs. T, ex Ic, nFS. [♂ **2**]

Quail *Coturnix coturnix* PHASIANIDAE L 17–19 cm. Seldom seen, except when flushed; like tiny partridge, but sandier, head stripy, upperparts and flanks with whitish streaks; ♂ throat whitish, divided and outlined with black-brown, ♀ breast spotted. Unmistakable call of ♂ liquid 'quic, quic-ic' (or 'wet-me-lips'), by night and day. Grasslands, clover, young corn. Nest hollow sparsely lined grasses, in rough grass, low crops, 7–12 eggs, May–Sep. Food grasses, clovers, weed seeds, insects. Summer sBr, eIr, Fr, Lu, Be, Ne, De, Ge, Cz, Po, erratically north to nBr, nIr, sFS. [♂ **3**] [♀ **4**]

Pheasant *Phasianus colchicus* PHASIANIDAE L ♂ 75–90 cm, ♀ 52–64 cm. Large, tail long (♂ <50 cm) and pointed; ♂ green head, red wattles, coppery body and tail; ♀ brown, mottled and barred buff and blackish. Main call of ♂ far-carrying 'korrk-korrk'. Wood edges, young plantations, scrub, arable fields, parkland, reed-beds. Nest hollow lined grasses, lvs, under cover, 7–15 eggs, Apr–Jul. Food lvs of many plants, seeds, grain, berries, roots, insects, earthworms, also slugs, lizards, voles. (T, ex Ic, all but sFS, from Asia.) [♂ **5**] [♀ **6**]

Corncrake *Crex crex* RALLIDAE L 25–28 cm. Seldom seen, usually heard; like small gamebird with short neck, tapering body; bright chestnut on wings in flight. Call of ♂ persistent double-note ('crex-crex'), like wood rasped along comb; most often at night, also by day. Grasslands, also clover, corn, damp ground with rushes. Nest hollow lined dead grass, in mowing grass, nettles, other rank vegetation, 8–12 eggs, May–Aug. Food mainly insects, also earthworms, slugs, snails, seeds. Summer T, ex Ic, but only nBr, sFS, often local. [**7**]

Little Bustard *Tetrax tetrax* OTIDIDAE L 41–45 cm. Small for bustard, like large gamebird with long neck, legs; breeding ♂ has blue-grey face, bold black-and-white neck-pattern; ♀ mainly buff with black markings; in flight, wings white with black tips. Difficult to see on ground, where may lie flat. Display note of ♂ like distant snort of horse. Grasslands, cornfields, clover. Nest hollow unlined, on ground in low vegetation or open, 3–4 eggs, May–Jul. Food clovers, grass, other lvs, also larger insects, snails, worms, frogs, voles. Summer Fr. [♂ **8**] [♀ **9**]

Great Bustard *Otis tarda* OTIDIDAE L ♂ 95–107 cm, ♀ 76–84 cm. Large, body stout, neck and legs long and thick; ♂ has moustache of long, whitish bristles, chestnut breast-band; in flight, wings largely white with spread black tips. Shy. Displaying ♂ blows up throat-pouch, turns himself 'inside out' in mass of white feathers. Barking grunt when breeding. Grassy plains, rolling cornlands. Nest hollow unlined but for trampled stalks, in long grass, corn, 2–3 eggs, May–Jul. Food cereals, other crops, grasses, clovers, weeds, seeds, also insects, voles, other animals. eGe, Cz, wPo, but scarce; wanders west in winter. [♂ **10**] [♀ **11**]

10
11
9
8
1
2
7
6
5
4
3

Stone-curlew *Burhinus oedicnemus* BURHINIDAE L 39–42 cm. Large, ungainly, head rounded; streaked sandy-brown with two bold, black-edged, whitish wing-bars; yellow bill (black-tipped), eyes, legs yellow. Runs furtively hunched. Curlew-like 'coor-lee' in flight; prolonged wailing on ground. Chalk downs, stony plains, heaths, rides in open woodland, locally on arable, sand-dunes. Nest scrape lined pebbles, rabbit droppings, usually on bare ground, 2 eggs, Apr–Aug. Food molluscs, earthworms, insects, also voles. Mainly summer seBr, Fr, eGe, Cz, Po; some winter WE. **[1]**

Lapwing *Vanellus vanellus* CHARADRIIDAE L 29–32 cm. Looks pied, with long crest; metallic green back, undertail red-buff; throat white in winter; in flight, black-ended white tail, black-and-white undersides of rounded, slow-flapped wings. Display flight erratic, tumbling. Familiar 'pees-weet'. Open country, esp arable, pastures, rushy fields, gravel pits, moors; winters fields, marshes, estuaries. Nest scrape lined grass (thickly in wet sites), on open ground, 3–4 eggs, Mar–Jul. Food insects, spiders, slugs, snails; earthworms, also seeds, grass lvs, cereals. T, ex Ic, nFS, but largely summer Cz, Po, FS. [summer **2**] [winter **3**]

Woodcock *Scolopax rusticola* SCOLOPACIDAE L 33–36 cm. Stout, bill long; head with broad black bars, upperparts marbled, underparts barred. Well camouflaged, rises with twisting flight. Rather silent, except thin 'tsiwick' and low croaking in 'roding' display-flight. Open woods with wetter areas for feeding; also scrub, young plantations, open bracken and heather. Nest scrape lined dead lvs, under light cover of dead bracken, brambles, often close to tree, 4 eggs, Mar–Jun. Food earthworms, insects, spiders, slugs, snails, seeds. T, ex Ic, nFS, but only summer Ge, Cz, Po, Fi, all but swSC. **[4]**

Stock Dove *Columba oenas* COLUMBIDAE L 32–34 cm. Smaller, darker, more compact than woodpigeon, without white on wings, neck; 2 short black bars on inner wings. Distinctive 'ooo-woo-oo-up', deep, gruff, urgent. Old woods, parks, cliffs, ruins, feeding in fields. Nest of sticks, roots, lvs, sometimes nil, in hole in tree, thatch, rocks, rabbit burrow, old magpie or pigeon nest, 2 eggs, Mar–Sep. Food grain, clovers, turnip lvs, other plants, also small snails, insect larvae. T, ex Ic, nFS, but only summer De, eGe, Cz, Po, FS. **[5] [6]**

Woodpigeon *Columba palumbus* COLUMBIDAE L 39–42 cm. Large, heavily built; bold white wing-bar, white patch on side of neck. 'Explodes' noisily from trees; often in flocks. Familiar, sleepy 'cooo-coo, coo-coo, cu'. Woods, locally in towns. Nest platform of twigs, lined roots, grasses, in bush or tree, on ground in nettles, heather, marram, 2 eggs, Mar–Nov. Food cereals, clovers, roots, lvs and frs of many crops and other plants, also earthworms, snails, slugs, insect larvae. T, ex Ic, nFS, but only summer eGe, Cz, Po, Fi, all but swSC. **[7] [8]**

Collared Dove *Streptopelia decaocto* COLUMBIDAE L 30–33 cm. Distinguished from turtle dove by longer, white-cornered tail, white-edged black half-collar, uniform grey-brown upperparts. Parties feed together. Characteristic 'coo-cooo-cuh', accented on 2nd syllable; nasal flight-call 'kwurr'. Farms, villages, urban parks, gardens. Nest platform of grass stems, sometimes twigs, in tree, esp conifer, rarely on building, 2 eggs, Feb–Oct. Food grain, esp in chicken-runs, also seeds, berries, some scraps. T, ex Ic (has bred), Fi, nSC. **[9] [10]**

Turtle Dove *Streptopelia turtur* COLUMBIDAE L 27–29 cm. Small, slender, with graduated, white-edged, black tail; black-and-white patch on side of neck, upperparts red-brown mottled black; juv duller, no neck-patch. Sleepy, purring 'rroorrrr', repeated 2–5 times. Thick hedges, downs, commons with thorns, copses, wood-edges, feeding in open fields. Nest platform of fine twigs, lined roots, grass, in bush, young tree, 2 eggs, May–Aug. Food fumitory, other plant seeds and lvs. Summer sBr, Fr, Lu, Be, Ne, Ge, Cz, Po, rare Ir, De. **[11] [12]**

1
2
3
4
5
6
7
8
9
10
11
12

Cuckoo *Cuculus canorus* CUCULIDAE L 32–34 cm. Tail long, graduated, wings pointed; upperparts and throat blue-grey, underparts whitish barred dark grey; juv has white nape-patch, strongly barred red-brown or faintly marked grey-brown upperparts. Familiar call of ♂ far-carrying 'cuc-coo'; ♀'s bubbling chuckle less well-known. Open woods, bushy commons, farmland with hedgerows, also moors, tundra, sand-dunes. ♀ lays in nests of other spp, esp pipits, wagtails, warblers, dunnock, robin, specializing on one host, 6–18 eggs during May–Jul. Food mainly insects, esp larvae; ♀ eats one of host's eggs. Summer T, ex Ic. [ad **1**] [red juv **2**]

Barn Owl *Tyto alba* TYTONIDAE L 33–36 cm. Ghostly shape in dusk or headlights; upperparts golden-buff mottled grey and white, underparts white (Br, Ir, wFr) to rich buff spotted with dark brown. Wavering, buoyant flight. Farm buildings, ruins, parks, hedgerows with old trees, also cliffs, quarries, hunting over open country. Nest scrape in debris, in roof-space, hole in thatch or straw, dovecote, hollow tree, rock crevice, 4–7 eggs, Mar–Oct. Food small mammals, birds, also moths, frogs. T, ex Ic, Fi, No, but only extreme sSw. [light **3**] [dark **4**]

Scops Owl *Otus scops* STRIGIDAE L 18–20 cm. Smallest European owl with ear-tufts (often inconspicuous), slimmer, more tapered than little owl; grey-brown or red-brown finely streaked and barred black-brown. Nocturnal; roosts by day in tree. Soft, musical, whistling 'pew' persistently repeated, confusable with bell-like voice of midwife toad. Open woods, parks, gardens, roadside trees. Nest scrape in debris in hole in tree, wall, bank, old nest of other bird, 3–6 eggs, Apr–Jul. Food mainly large insects. Summer Fr, sCz. [**5**]

Little Owl *Athene noctua* STRIGIDAE L 21–24 cm. Small, compact, squat, flat-headed; yellow eyes and flattened facial discs give fierce, frowning expression. Crepuscular, partly diurnal, often seen on telegraph-pole; bounding flight. Plaintive 'keeu'. Farmlands with old hedgerow trees, parks, orchards, wood-edges, quarries, also

screes, cliffs, sand-dunes. Nest scrape in debris in hole in tree, building, haystack, quarry, rabbit burrow, 3–5 eggs, Mar–Jul. Food insects, small mammals, birds, also slow-worms, earthworms, other invertebrates. T, ex nBr, Ir, Ic, FS, (but sBr). [**6**]

Tawny Owl *Strix aluco* STRIGIDAE L 37–39 cm. Head large, rounded, body portly; eyes black in grey-brown facial discs; red-brown to grey, mottled and streaked light and dark brown, whitish patches on shoulders. Nocturnal. Sharp 'kewick'; familiar hooting 'hoo-hoo-hoo' ends with long-drawn, tremulous 'hoooooooooo'. Woods, farmlands, parks, churchyards, gardens, wherever old trees. Nest scrape in debris in hole in tree, building, rock-crevice, sometimes in old crow or raptor nest or on ground, 2–5 eggs, Mar–Jun. Food mainly small mammals, birds, also fish, frogs, earthworms, large insects. T, ex Ir, Ic, nFS. [red **7**] [grey **8**]

Roller *Coracias garrulus* CORACIIDAE L 30–32 cm. Crow-like, with powerful bill; mainly greenish-blue with chestnut back; in flight, wings show striking pattern of vivid blue, turquoise, purplish-blue and black; juv duller, browner, less blue. Loud, harsh 'krak-ak' and chattering. Shrike-like pouncing from perch on prey; tumbling display-flight. Woods, open country with old trees. Nest in hole in tree, wall, cliff, 4–5 eggs, May–Jul. Food insects on ground, esp beetles, also lizards, centipedes, fruit. Summer Ge, Cz, Po, sSw (Gotland), but scarce. [**9**]

Hoopoe *Upupa epops* UPUPIDAE L 27–29 cm. Bill long, curved; black-tipped crest forms point at rear or fan over head; body pinkish, black-and-white wings, tail. Flapping flight with rounded wings almost closed at each beat. Distinctive, low, but far-carrying 'poo-poo-poo'. Wood-edges, parks, farmlands with old hedgerows, tree-lined rivers, orchards. Nest hollow in hole in tree, wall, pile of stones, 5–8 eggs, May-Jul. Food insect larvae, large insects, also other invertebrates, small lizards. Summer Fr, Lu, Ge, Cz, Po, rare Be; passage also Br, Ir, Ne, sFi, sSw. [**10**]

Wryneck *Jynx torquilla* PICIDAE L 16–17 cm. Quite unlike other woodpeckers, with small bill, slim shape, long rounded tail; upperparts grey-brown delicately marked with light and dark, underparts barred on throat, arrow-marked below. Clings like woodpecker to trunks; hops on ground with raised tail. Shrill 'quee-quee-quee' (*cf* lesser spotted). Open woods, parkland, orchards, hedgerows with old trees. Nest hollow in debris in hole in tree, bank, wall, thatch, 7–10 eggs, May–Aug. Food chiefly ants, also beetles, moths, larvae, spiders, woodlice. Summer Fr, Lu, Be, Ne, De, Ge, Cz, Po, sFS, rare seBr, nBr. **[1]**

Grey-headed Woodpecker *Picus canus* PICIDAE L 24–26 cm. Like green woodpecker, but smaller, head and underparts grey with narrow black moustache, no red on crown, only ♂ has red forehead; juv browner, with brown-barred flanks. 'Laughing' call less harsh, deeper, slower. Woods, scattered trees, lowlands to broadleaved mountain forests. Nest hole excavated in tree, unlined, 4–5 eggs, Apr–Jul. Food mainly wood-boring larvae, ants, some seeds, berries. Fr, Lu, Ge, Cz, Po, sFS. [♂ **2**] [♀ **3**]

Green Woodpecker *Picus viridis* PICIDAE L 31–33 cm. Crown red with black on sides of face extending into moustache (broader, red-centred on ♂), upperparts dull green with yellow rump, underparts pale grey-green; juv paler, with light spots above, blackish streaks and bars below. Loud, laughing 'keu-keu-keu'. All woodpeckers use short, stiff tails as props in climbing trunks by jerky hops, chisel bills for tapping and boring, long tongues for extracting insects. Timbered hedgerows, parks, copses, open woodlands, wood-edges. Nest hole excavated in tree, 5–7 eggs, Apr–Jul. Food wood-boring larvae, ants and pupae, also acorns, seeds. T, ex nBr, Ir, Ic, Fi, nFS. [♂ **4**] [♀ **5**]

Great Spotted Woodpecker *Dendrocopos major* PICIDAE L 22–24 cm. Boldly pied; black crown, black cheek-bar joining nape, black back, white shoulder-patches, red undertail-coverts; ♂ has patch of red on nape; juv whole crown red. Loud, sharp 'tchick', sometimes speeded into chatter; often drums on dead wood. Woodlands, esp broadleaved, also parks, orchards, gardens, seldom in hedgerow timber. Nest hole excavated in tree, 4–7 eggs, May–Jul. Food wood-boring larvae, also spiders, nestling birds, beechmast, nuts, seeds. T, ex Ir, Ic, nFS. [♂ **6**] [♀ **7**]

Middle Spotted Woodpecker *Dendrocopos medius* PICIDAE L 21–22·5 cm. Boldly pied; red crown without black edge, broken moustache and cheek-band (head looks pale), black back, white shoulder-patches, pink (not red) belly and undertail, streaked flanks; ♀ duller, with paler red crown. Chatter like great spotted, but first note higher; slow, nasal 'wait, wait' in spring; drums rarely. Broadleaved woodland, esp hornbeam, beech, oak. Nest hole excavated in tree, 4–8 eggs, Apr–Jul. Food wood-boring larvae, beechmast. Fr, Lu, Ge, Cz, Po, irregular Ne. [♂ **8**]

White-backed Woodpecker *Dendrocopos leucotos* PICIDAE L 25–26 cm. Pied; black shoulders, white or barred back and rump, barred wings, streaked flanks, pink undertail-coverts, black moustache but broken cheek-band; ♂ forehead whitish, crown red, black-edged, ♀ crown black, juv traces of red. Calls rather little, more quietly than great spotted. Woods, esp broadleaved in hilly country; round towns in winter. Nest hole excavated in tree, 4–5 eggs, May–Jul. Food wood-boring larvae, also acorns, nuts, berries. sGe, eCz, ePo, sFS. [♂ **9**] [♀ **10**]

Lesser Spotted Woodpecker *Dendrocopos minor* PICIDAE L 14–15 cm. Pied; sparrow size, barred back and wings, no red on undertail; ♂ crown red, black-edged, ♀ whitish, juv some red. Loud, shrill 'pee-pee-pee', weaker, less musical than wryneck; commonly drums. Open woodlands, copses, parks, hedgerow trees, avenues, old orchards. Nest hole excavated in tree, often underside of sloping branch, 4–6 eggs, Apr–Jul. Food wood-boring larvae, spiders, occasionally berries. T, ex nBr, Ir, Ic. [♂ **11**] [♀ **12**]

1
2
3
4
5
6
7
8
9
10
11
12

Swift *Apus apus* APODIDAE L 16–17 cm. Scythe-shaped wings and short, forked tail; all sooty-black except pale throat. Distinctive, rapid beats of stiff wings. Noisy when breeding, screaming parties chasing round buildings. Exclusively aerial, over urban areas, any open country, often over water. Social; nest of aerially collected plant fragments, feathers, in hole in building, cliff, tree, 2–3 eggs, May–Aug. Food insects, spiders, caught in air. Summer T, ex Ic, nSC. [**1**]

Short-toed Lark *Calandrella brachydactyla* ALAUDIDAE L 13·5–14·5 cm. Small, no crest, bill short, pointed, finch-like; sandy-buff with reddish cap above, nearly white below, dark mark at side of breast; juv speckled above, streaked across breast. Short, hard 'tchi-chirrp'; short song-phrase constantly repeated in yoyo flight. Sandy cultivation, dunes, dry salt-marshes. Nest of dry grasses, roots, lined hair, wool, feathers, in hollow on ground, usually sheltered by plant, 3–5 eggs, May–Jul. Food small seeds, also insects. Summer wFr, but very local. [**2**]

Crested Lark *Galerida cristata* ALAUDIDAE L 16·5–17·5 cm. Plump, with long crest, stout and slightly decurved bill, short tail; upperparts less streaked than skylark, orange-buff underwings, tail buff at sides; juv more spotted, with shorter crest. Liquid, whistling 'whee-wheeoo'; song shorter, less musical than skylark, not in high flight. Open country, esp arable, dry sandy places, also roadsides, railways, wasteland. Nest of grasses, sometimes lined hair, in hollow on ground, 3–5 eggs, Apr–Jul. Food grain, many seeds, also insects, larvae. Fr, Lu, Be, Ne, De, Ge, Cz, Po, sSC. [ad **3**] [juv **4**]

Woodlark *Lullula arborea* ALAUDIDAE L 14·5–15·5 cm. Bill rather fine, crest rounded but often inconspicuous, tail short with white tips (not sides); pale superciliaries meeting on nape, white-edged black mark at bend of wing; juv more spotted above. Liquid 'titlooeet'; song very sweet, musical, characteristic phrase being rich 'lu-lu-lu'. Open country with scattered trees, commons, parkland, wood-edges, esp sandy soils. Nest of

grasses, moss, lined fine grass, hair, on ground, often half-hidden by plants, 3–4 eggs, Mar–Aug. Food mainly insects, larvae, spiders, also seeds in autumn. T, ex nBr, Ir, Ic, nFS, but only summer Gc, Cz, Po, sFS. [**5**]

Skylark *Alauda arvensis* ALAUDIDAE
L 17–18 cm. Crest rounded but often prominent, tail rather long, bill fairly stout; upperparts brown, streaked blackish, underparts buff-white, well-streaked on breast; white tail-sides, whitish hind-margin of wings conspicuous in flight; juv more spotted with less crest. Rippling 'chirrup'; musical song sustained for minutes on end high in air. Open grasslands, cultivation, marshes, sand-dunes, moors. Nest of grasses, sometimes lined hair, on ground in grass or crops, 3–5 eggs, Apr–Aug. Food grain, many seeds, some lvs, earthworms, insects, spiders. T, ex Ic, nSC, but mainly summer Cz, Po, FS. [ad **6**] [juv **7**]

Swallow *Hirundo rustica* HIRUNDINIDAE
L 17–19 cm. Slender, with long wings, tail-streamers; upperparts and breast-band dark blue, forehead and throat chestnut-red, underparts creamy-white; juv duller, with shorter streamers. Twittering calls and song, latter mixed with trilling warble. Largely aerial over open country, esp arable and grass round farm buildings, often concentrated over water. Cup nest of mud and straw, lined grass, feathers, on rafter, in shed, under bridge, 4–6 eggs, May–Sep. Food insects, caught on wing. Summer T, ex Ic, nFS. [ad **8**] [juv **9**]

House Martin *Delichon urbica*
HIRUNDINIDAE L 12–13 cm. Tail slightly forked; upperparts blue-black, rump and underparts white. Calls more chirruping than swallow; infrequent song prolonged twittering. Largely aerial over open country, esp cultivation near farms, as well as towns, villages, locally cliffs, often concentrated over water. Colonial; nest half-cup of mud mixed with plant fibres, lined fine grass, feathers, under eaves, bridge, cliff overhang, 4–5 eggs, May–Oct. Food insects, caught on wing. Summer T, ex Ic. [**10**]

Tawny Pipit *Anthus campestris* MOTACILLIDAE L 16–17 cm. Large, slim, wagtail-like pipit with longish legs; sandy above, paler below, almost unmarked but for creamy superciliaries, row of dark spots on wings, buff-white tail-edges; juv upperparts darker, marked pale buff, breast streaked. Wagtail-like 'tsweep'; song 'chivee-chivee' in high flight. Sandy wastes or cultivation, dry grassland, dunes. Cup nest of grass, weeds, roots, lined finer grass, hair, on ground by tussock, 4–5 eggs, May–Jul. Food insects. Fr, Lu, Be, Ne, De, Ge, Cz, Po, sSw, but generally scarce/local. [ad **1**] [juv **2**]

Tree Pipit *Anthus trivialis* MOTACILLIDAE L 15–15·5 cm. White tail-edges; confusable with meadow pipit *A. pratensis*, but plumper, stouter bill, more lightly streaked on yellower breast, pink legs. Habitually on trees. Distinctive hoarse 'teez'; song loud, musical, chaffinch-like notes ending in descending 'seea-seea-seea' during 'parachute drop' from upward flight. Open ground or scrub with scattered trees, clearings, young plantations, heaths, bracken slopes. Cup nest of grass, on moss base, lined finer grass, hair, in bank, side of tussock, under brambles, bracken, 4–6 eggs, May–Aug. Food insects. Summer T, ex Ir, Ic. [**3**]

White/Pied Wagtail *Motacilla alba* MOTACILLIDAE L 17–18 cm. Tail long, constantly moved up and down; ♂ boldly black, grey, white; ♀ greyer, less black on head, breast; juv blackish, grey-brown, off-white. Pied ssp (Br, Ir, adjacent coasts) has black rump, ♂ also black back, where white ssp grey. Characteristic 'tchizzik'; song babbling twitter of slurred notes. Often by water, in open country, farmland, also towns, quarries, hill streams. Cup nest of moss, grass, twigs, roots, lined hair, feathers, wool, in recess in wall, rock, bank, haystack, ivy, tree, old nest, Apr–Aug. Food mainly insects. T, but only summer Ic, Ne, De, Ge, Cz, Po, FS. [♂ white **4**] [♂ pied **5**] [♀ **6**] [juv **7**]

Wren *Troglodytes troglodytes* TROGLODYTIDAE L *c*9·5 cm. Tiny, stumpy; tail short, cocked; red-brown, closely-barred

blackish, with paler superciliaries, underparts. Hard ticking or churring; song loud, clear phrase of shrill notes with final trill. Wherever trees, hedgerows, scrub, banks, walls, esp farmland, also hills, coasts. Domed nest of dead lvs, bracken, moss, grass, lined feathers, with side-entrance, in any fork, cranny, hollow, 5–6 eggs, Apr–Aug (♂ builds several, ♀ lines 1). Food insects, spiders, seeds. T, ex nFS, but only summer Fi. [**8**]

Dunnock *Prunella modularis* PRUNELLIDAE L 14–15 cm. Bill slender; red-brown streaked blackish, with greyer head, breast; juv less red, more spotted, with browner head. Unobtrusive, feeding on ground with shuffling gait, flicking wings. Piping 'tseep'; song clear, high jingle. Wood-edges, hedgerows, bushy commons, gardens, moorland scrub. Cup nest of moss, lvs, roots, grass, on base of twigs, lined moss, wool, hair, feathers, in bush, hedge, low tangle, ivy, bank, 4–5 eggs, Apr–Aug. Food insects, spiders, seeds. T, ex Ic, but only summer ePo, Fi, all but swSC. [**9**]

Thrush Nightingale *Luscinia luscinia* TURDIDAE L 16–17 cm. Distinguished from nightingale by darker, olive-brown upperparts, darker red-brown tail, mottled breast. Song more powerful, with purer, bell-like notes, but lacks rising crescendo. Damp woods, esp alder, birch, with undergrowth, waterside thickets. Nest and food as nightingale. Summer eDe, nGe, eCz, nePo, sFS. [**10**]

Nightingale *Luscinia megarhynchos* TURDIDAE L 16–17 cm. Upperparts red-brown with redder tail, underparts grey-brown to whitish; juv spotted, with chestnut tail. Notes include liquid 'wheet', hard 'tac', harsh 'krrr'; song rich, loud, repetitive phrases, bubbling 'jug-jug-jug', slow 'piu, piu, piu' rising to crescendo, hard trills, day and night. Woods with thick undergrowth, thickets, tangled hedgerows. Cup nest of lvs, grass, lined hair, on ground under low vegetation or more open, sometimes on stump or in hedge, 4–5 eggs, May–Jul. Food insects, earthworms, berries. Summer sBr, Fr, Lu, Be, Ne, Ge, Cz, swPo. [**11**]

Robin *Erithacus rubecula* TURDIDAE L 13·5–14·5 cm. Plump, seemingly neckless; face and breast red-orange, edged pale grey; juv mottled and spotted brown and buff. Scolding 'tic-tic', thin plaintive 'tsweee'; spring song (♂) of warbling trills, prolonged notes, autumn song (♂ ♀) slower, more melancholy. Woods; tameness near houses mainly in Br, Ir. Nest of moss, some grass, often on dead lvs, lined roots, hair, in bank, tree-hole, ivy, wall, shed, 5–7 eggs, Mar–Aug. Food insects, earthworms, seeds, berries. T, ex Ic (annual vagrant), nFS, but mainly summer sFS. [ad **1**] [juv **2**]

Redstart *Phoenicurus phoenicurus* TURDIDAE L 13·5–14·5 cm. Rump and constantly-quivered tail chestnut-red at all ages; ♂ grey, orange, black, white; ♀ grey-brown above, orange-buff below; juv mottled like young robin. Plaintive 'wheet', scolding 'whee-tic-tic'; song short warble ending in jingle. Broadleaved woods, old pine forests, parks, orchards, heaths. Nest of grass, bark, roots, moss, lined hair, feathers, in hole in tree, wall, building, rocks, 5–7 eggs, May–Aug. Food insects, also spiders, earthworms, berries. Summer T, ex Ic, but rare Ir. [♂ **3**] [♀ **4**]

Blackbird *Turdus merula* TURDIDAE L 24–26 cm. ♂ all-black with orange-yellow bill, eye-ring; ♀ dark brown above, paler and more rufous below, with whitish throat, brown or yellowish bill; juv paler, more rufous, more mottled. Chattering alarm, also 'chink, chink', anxious 'chook'; song rich, fluty warbling. Woods, scrub, hedges, gardens. Cup nest of moss, grass, lined mud, then more grass, in bush, tree-fork or cavity, in or on wall or building, on ground under cover, 3–5 eggs, Mar–Aug. Food earthworms, insects, seeds, berries. T, ex Ic (has bred), nFS; winter also Ic. [♂ **5**] [♀ **6**]

Fieldfare *Turdus pilaris* TURDIDAE L 24·5–26·5 cm. Grey head and rump contrast with chestnut back, black tail; throat and breast rusty, streaked and arrow-marked with black; juv duller with light streaks on upperparts. Flight-call harsh 'chack-chack-chack'; feeble song of squeaky, whistling notes. Woods, parks, orchards, gardens; winters farmland with hedgerows, open woodland. Often colonial; bulky nest of grass with mud layer under inner lining, in tree, bank, 4–6 eggs, Apr–Aug. Food slugs, snails, insects, earthworms, berries. Summer eFr, Lu, De, Ge, Cz, Po, FS, rare nBr; winter T, ex nFS. [**7**]

Song Thrush *Turdus philomelos* TURDIDAE L 22–23·5 cm. Upperparts brown, breast and flanks yellow-buff with small blackish spots; underwing yellow-buff in flight; juv streaked buff above. Flight-call thin 'sip', alarm repeated 'chook'; loud song, repetition of clear phrases. Woods, copses, hedges, gardens. Cup nest of grass, moss, on twigs, lined mud, wood-pulp, dung, in bush, ivy, tree-fork, bank, 3–6 eggs, Mar–Aug. Food slugs, snails, earthworms, insects, seeds, berries. Summer T, ex Ic; winter Br, Ir, Fr, Lu, Be, Ne, sNo. [**8**]

Redwing *Turdus iliacus* TURDIDAE L 20–21·5 cm. Distinguished from song thrush by creamy superciliaries, reddish flanks and underwing; juv pale-streaked above. Flight call soft 'see-ip', alarm harsh 'chittick'; song repeated phrase of 3–6 fluty notes, varies locally. Woods of birch, alder, also young conifers, scrub; winters open country with hedgerows. Nest of grass, twigs, moss, mud, lined grass, in tree, bush, bank, or on ground, 4–6 eggs, Apr–Aug. Food slugs, snails, insects, seeds, berries. Summer Ic, FS, rare nBr, De, Po; winter T, ex Cz (passage), Po, Fi, all but swSC. [**9**]

Mistle Thrush *Turdus viscivorus* TURDIDAE L 26–27·5 cm. Larger than song thrush, greyer upperparts, whiter underparts with larger spots, also white underwing, whitish tail-corners; juv spotted white above. Churring rattle; song often confused with blackbird, but less varied repetition of 3–6 phrases. Woods, parks, farmland with trees, orchards. Cup nest of grass, plant stems, moss, wool, no mud, lined grass, in tree-fork, on wall ledge, 3–5 eggs, Feb–Jul. Food berries, seeds, slugs, snails, earthworms, insects. T, ex Ic, nwSC, but only summer Fi, most SC. [ad **10**] [juv **11**]

1
2
4
3
7
9
10
8
11
6
5

Icterine Warbler *Hippolais icterina* SYLVIIDAE L 13–13·5 cm. Peaked crown, pointed wings projecting beyond tail-base; olive-brown and yellow with yellowish wing-patch, blue-grey legs. Liquid 'deederoid'; song loud, sustained jumble of repeated musical and discordant notes. Woods, parks, thickets, gardens. Cup nest of grass, dead lvs, roots, decorated with wool, fibres, cobwebs, lined fine grass, hair, in bough-fork of small tree, 4–5 eggs, May–Jul. Food insects, larvae, spiders, berries. Summer eFr, Lu, Be, Ne, De, Ge, Cz, Po, sFi, sSw, wNo; scarce passage eBr, sIr. [**1**]

Melodious Warbler *Hippolais polyglotta* SYLVIIDAE L 12·5–13 cm. Smaller than icterine, rounded crown, shorter wings reaching only to tail-base; browner above, less clear wing-patch mainly in spring, deeper yellow below, browner legs. Sparrow-like chatter; song more musical, quicker, less repetitive than icterine, with fewer harsh notes, but equally sustained. Open woods, scrub, hedgerows, often near water. Cup nest like icterine, but usually lower, in bush, 3–5 eggs, May–Jul. Food as icterine. Summer Fr; scarce passage sBr, sIr. [**2**]

Barred Warbler *Sylvia nisoria* SYLVIIDAE L 15–15·5 cm. Bulky, with stout bill, yellow eyes, long white-edged tail, 2 whitish wing-bars; ♂ grey above, below whitish, barred with dark crescents; ♀ browner, less barred; juv buffish below, with few or no bars. Harsh chatter, often built into song which otherwise resembles garden warbler. Thickets, scrub, wood-edges, young plantations. Cup nest loosely built of grass, lined hair, roots, often in thorn bush, 5 eggs, May–Jul. Food as whitethroat. Summer sDe, eGe, Cz, Po, swFi, sSw; scarce passage eBr, sNo. [♂ **3**] [juv **4**]

Lesser Whitethroat *Sylvia curruca* SYLVIIDAE L 13–13·5 cm. Shorter tail than whitethroat, dark mask, greyer upper-parts, no chestnut on wings. Skulking, best located by song: loud rattle on one note recalling cirl bunting, but often preceded by low warble. Thorn hedges with brambles, wood-edges, young plantations, tangled gardens. Small cup nest of dead grass, with wool and spiders' cocoons on rim, lined fine grass, roots, hair, in thorn, bramble, small evergreen, 4–6 eggs, Apr–Aug. Food as whitethroat. Summer T, ex nBr, Ir, Ic, wFr, nFS. [**5**]

Whitethroat *Sylvia communis* SYLVIIDAE L *c*14 cm. Often raises crest and cocks longish, white-sided tail; white throat, rusty area on wings; summer ♂ has grey cap, pinkish breast, ♀ browner head, less pink (but some overlap). Harsh, scolding churr and quiet 'weet, weet, wit-wit-wit'; song brisk, chattering warble, often in 'dancing' flight. Tangled hedgerows, scrub, wood-edges, young plantations. Cup nest of dry grass, lined fine grass, rootlets, hair, low in bramble, grass-grown bush, 4–5 eggs, Apr–Aug. Food insects, larvae, spiders, berries. Summer T, ex Ic, nFS. [♂ **6**] [♀ **7**]

Garden Warbler *Sylvia borin* SYLVIIDAE L *c*14 cm. Plump, head round, bill stubby; brown above, pale buff below, featureless yet distinctive. Calls resemble blackcap, and song often confused, but quieter, faster, more even and sustained. Open woodland with undergrowth, scrub with or without trees, young plantations. Cup nest like whitethroat, but flatter, often where bramble stems cross among nettles, also in gorse, young conifer, generally low, 4–5 eggs, May–Aug. Food as whitethroat. Summer T, ex Ic, but local nBr, Ir, nFS. [**8**]

Blackcap *Sylvia atricapilla* SYLVIIDAE L *c*14 cm. Cap glossy black on ♂, red-brown on ♀; ♂ grey-brown above, grey below, ♀ browner. Scolding 'tack, tack', harsh churr; song short, clear warble with rich, pure notes, but rambling subsong recalls garden warbler. Open woods with thick undergrowth, also tree-lined lanes, orchards, shrubby gardens. Cup nest neat, compact, of dry grass with wool and cobwebs on rim, lined fine grass, roots, hair, in bush, small tree, ivy, 4–5 eggs, Apr–Aug. Food as whitethroat. Summer T, ex Ic, nFS, but local nBr, wIr; some winter Br, Ir, wFr. [♂ **9**] [♀ **10**]

1
2
3
4
8
6
5
7
9
10

Grasshopper Warbler *Locustella naevia* SYLVIIDAE L 12–13 cm. Upperparts olive-brown, often looking yellowish, streaked dark brown; tail rounded, faintly barred; underparts pale buff, usually lightly streaked on breast. Skulking, located by song: high-pitched, insect-like trill, recalling sound of fishing reel, of rapid double-notes at *c*1800/min, far-carrying $<\frac{1}{2}$ km (but inaudible to some ears). Dry or wet grassland with scrub, young plantations, bushy commons, open woods, to marshes, moors. Nest of grass, often on base of dead lvs, with finer lining; well-hidden on or near ground in tussock or clump of thick grass, bramble, gorse, sedge, rush, entered by run, 5–6 eggs, May–Aug. Food insects, larvae, also spiders, woodlice. Summer T, ex Ic, No, but only sFi, sSw. [**1**]

Greenish Warbler *Phylloscopus trochiloides* SYLVIIDAE L 10·5–11 cm. Distinguished from chiffchaff by short, whitish wing-bar, longer and bolder superciliaries, whitish-yellow underparts; wings short, legs dark grey-brown. Thin 'whee-e', second syllable lower and flatter; song powerful trill, usually preceded by rapid repetitions of call. Open woodland, wood-edges, also scrub. Domed nest of moss, grass, dead lvs, lined hair, feathers, with side-entrance, on or near ground in vegetation or crevice, 4–6 eggs, Jun–Jul. Food mainly insects, larvae. Summer nGe, nPo, seFi, has bred sSw. [**2**]

Bonelli's Warbler *Phylloscopus bonelli* SYLVIIDAE L 11–12 cm. Looks grey, esp on head, with pale grey-brown upperparts, whitish underparts; yellow at bend of wings, yellowish patches on wings and (often hard to see) rump. Plaintive 'hoo-eet', more disyllabic than willow warbler; trill slower than wood warbler (*cf* lesser whitethroat). Open woodland, lowlands to mountains, often on slopes. Domed nest of grass, lined finer grass, hair, with side-entrance, on ground or in bank among low vegetation, 4–6 eggs, May–Jul. Food insects, larvae. Summer Fr, sGe, ?Lu. [**3**]

Wood Warbler *Phylloscopus sibilatrix* SYLVIIDAE L 12·5–13 cm. Larger, brighter

than other *Phylloscopus*, with long wings, shortish tail; throat and breast bright yellow, contrasting with white underparts. Call piping 'piu'; 2 songs: (1) accelerating notes ending in shivering trill, (2) falling series of piping whistles. Tall woods, esp beech, oak, ash, birch, with little undergrowth; scrub oak and birch on hillsides. Domed nest of grass, bracken fibres, moss, some lvs, lined fine grass, on ground where herbage low or absent, 5–7 eggs, May–Jul. Food mainly insects, larvae. Summer Br, Fr, Lu, Be, Ne, De, Ge, Cz, Po, sFS, rare Ir. [ad **4**] [juv **5**]

Chiffchaff *Phylloscopus collybita*
SYLVIIDAE L 10·5–11 cm. Shorter wings than willow warbler, clearer pale crescents above and below eyes, blackish feet; upperparts olive-brown (greyer in NE), underparts buff-white often with yellow at sides of breast. Wings and tail constantly flicked. Call soft 'hooeet'; song repetition of 'chiff' and 'chaff' in irregular sequence. Woods, copses, often with conifers, also tall scrub with scattered trees. Flattened domed nest of dead lvs, grass, bark, moss, lined feathers, with side-entrance, usually above ground in dense vegetation, 4–8 eggs, Apr–Aug. Food insects and larvae, spiders and eggs. Summer T, ex Ic, nFS, sSw; passage Ic, sSw; some winter sBr, sIr, wFr.
[WE, CE **6**] [NE **7**]

Willow Warbler *Phylloscopus trochilus*
SYLVIIDAE L 10·5–11 cm. Longer wings than chiffchaff, yellow superciliaries well-defined behind eyes, usually yellower underparts, paler feet. Plaintive 'hooeet'; song liquid, rippling phrase, beginning quietly, becoming louder and more emphatic, dropping away to faint ending. Wide range of open woodland, copses, bushy scrub, young plantations, lowlands to mountains; on passage in hedges, gardens. Spherical domed nest of moss, grass, bracken, lined grass, usually feathers, with underhung side-entrance, on or near ground in dense or sparse vegetation, 6–7 eggs, May–Aug. Food insects, larvae, spiders, also small earthworms, berries. Summer T, ex Ic; passage Ic. [ad **8**] [juv **9**]

Spotted Flycatcher *Muscicapa striata* MUSCICAPIDAE L 13·5–14·5 cm. Grey-brown, whiter below, with spotted crown, streaked breast. Sits on twig, sallies forth to snap insect, returns to perch. Thin, scratchy, robin-like 'tzee'; song squeaky notes. Wood-edges, parks, avenues, orchards, gardens. Cup nest of moss, bark, grass, lined roots, hair, wool, feathers, adorned cobwebs, lichens, against trunk on twigs or ivy, in open cavity, old thrush nest, or on stone ledge, 4–5 eggs, May–Aug. Food insects. Summer T, ex Ic. [1]

Red-breasted Flycatcher *Ficedula parva* MUSCICAPIDAE L 11–11·5 cm. Grey-brown with white patches at base of blackish, oft-cocked tail, underparts whitish-buff; ♂ always grey head, orange-red throat. Wren-like chatter; song bell-like and fluty notes, accelerating, then falling away. Old woodland, esp hornbeam, beech. Cup nest of moss, grass, plant down, lichens, lined hair, on twigs at side of trunk or in hollow, 5–6 eggs, May–Aug. Food insects. Summer Ge, Cz, Po, sFi, rare De, sSw; rare passage WE. [♂ 2] [♀ 3]

Collared Flycatcher *Ficedula albicollis* MUSCICAPIDAE L 12·5–13 cm. Summer ♂ distinguished from ♂ pied flycatcher by white collar, whitish rump, more white on forehead and wings, less at sides of tail; ♀ and autumn ♂ very like pied flycatcher, but greyer, sometimes with indication of collar or whitish rump. Long-drawn 'zeeb'; song shorter, simpler, slower than pied flycatcher. Habitat and nest as pied flycatcher, 6–7 eggs, May–Jul. Food insects. Summer eFr, sGe, Cz, sPo. [summer ♂ 4] [♀ 5]

Pied Flycatcher *Ficedula hypoleuca* MUSCICAPIDAE L 12·5–13 cm. Summer ♂ has black upperparts, white forehead, wing-patch, sides of tail, underparts; like ♀ in autumn, but forehead still whitish; ♀ olive-brown above, white in wings and tail. When flycatching, seldom returns to twig, often drops to ground. Chaffinch-like 'whit, whit'; song repeats sharp 'zy-et' 3–4 times, then liquid trill. Open woodland, esp oak, birch, parks, tree-lined rivers, orchards. Nest of lvs, grass, bark, roots, moss, wool, in hole in tree,

wall or building, nestbox, 5–8 eggs, May–Jul. Food insects. Summer T, ex seBr, Ir, Ic, rare nwFr, Be, Ne; passage T, ex Ic. [summer ♂ 6] [♀ 7]

Red-backed Shrike *Lanius collurio*
LANIIDAE L 16·5–17·5 cm. ♂ has black mask, blue-grey head and rump, chestnut back, white-edged black tail, pink-white underparts; ♀ red-brown above, buff below, with dark crescent markings; juv like ♀, but more heavily 'scaled'. Harsh 'chack, chack'; song jerky warbling, with mimicry. Scrub, rank hedgerows, gorse heaths. Bulky cup nest of moss, stems, lined fine grass, roots, hair, wool, in thorn bush, thick shrub, brambles, 3–6 eggs, May–Jul. Food insects, small birds and mammals, also frogs, lizards, earthworms, sometimes impaled on thorns in 'larders'. Summer Fr, Lu, Be, Ne, De, Ge, Cz, Po, sFS, now rare Br. [♂ **8**] [♀ **9**]

Great Grey Shrike *Lanius excubitor*
LANIIDAE L 23–25 cm. Tail long, graduated; grey above, with black mask, white superciliaries, white-marked black wings and tail; ♀ often has faint crescent marks on underparts; juv grey-brown, distinctly barred below. Harsh 'sheck, sheck', prolonged into chatter; song mixture of warbling, harsh notes, mimicry. Wood-edges, open woodland, orchards, scrub, heaths. Bulky cup nest of grass, moss, on base of twigs, lined roots, wool, hair, feathers, in thorn bush or high in tree, 5–7 eggs, Apr–Aug. Food small birds, some mammals, insects, also frogs, lizards. T, ex Br, Ir, Ic, sFS, but only summer nFS; winter also eBr. [♂ **10**]

Woodchat Shrike *Lanius senator*
LANIIDAE L 17–18 cm. Upperparts black, chestnut, white, ♀ duller; juv paler, greyer than juv red-backed shrike, more barred above, with light shoulders. Calls like other shrikes, but song richer, more varied warbling. Open country with scattered trees, wood-edges, orchards. Bulky cup nest of soft plants, roots, lined wool, hair, feathers, usually on outer branch of tree, 5–6 eggs, May–Jul. Food small birds, insects. Summer Fr, Lu, wGe, wCz, rare or irregular Be, Ne, sPo. [♂ **11**] [juv **12**]

Long-tailed Tit *Aegithalos caudatus* AEGITHALIDAE L *c*14 cm. Body tiny, tail long (*c*7·5 cm); pink, black and white, with broad black band over eye (WE, CE) or whole head white (NE); juv dark cheeks, no pink. Low 'tupp', rippling 'tsirrup'. Hedgerows, thickets, wood-edges. Ball nest of moss, bound cobwebs, hair, lined feathers (often 2000+), entrance near top, low in thick bush or high in tree, 7–12 eggs, Mar–Jul. Food insects. T, ex Ic, nFS. [WE **1**] [NE **2**]

Marsh Tit *Parus palustris* PARIDAE L *c*11·5 cm. Brown and whitish, with glossy black crown, small black bib; juv crown sooty like willow tit. Distinctive 'pitchew'. Woods, copses, orchards, timbered hedgerows. Nest of moss with felted cup of hair, plant down, in hole in tree, 6–8 eggs, Apr–Jun. Food insects, also seeds, berries. T, ex nBr, Ir, Ic, Fi, nFS. [**3**]

Willow Tit *Parus montanus* PARIDAE L 11·5–12 cm. Like marsh tit, but dull sooty crown, usually larger bib, pale patch on wing (more marked in nFS, where upperparts also greyer, cheeks whiter). Buzzing 'eez-eez-eez', nasal 'tchay-tchay-tchay'. Habitat overlaps marsh tit, but often wetter. Nest less moss, more grass, wood fibres, in hole excavated by ♀ in rotten stump, 6–9 eggs, Apr–Aug. Food insects, spiders, also berries. T, ex nBr, Ir, Ic, wFr, De. [WE **4**] [nFS **5**]

Coal Tit *Parus ater* PARIDAE L 10·5–11 cm. Head black, nape and cheeks white (yellower in Ir), 2 wing-bars; juv yellower. Clear, piping notes; song 'weechoo-weechoo', less strident than great tit. Woods, gardens, esp with conifers, to upland scrub. Nest of moss with felted cup, in hole in stump, wall, bank, ground, 7–11 eggs, Apr–Jul. Food insects, also spiders, seeds, beechmast, scraps. T, ex Ic, nFS. [ad **6**] [juv **7**]

Blue Tit *Parus caeruleus* PARIDAE L *c*11·5 cm. Crown, wings, tail bright blue, cheeks white, underparts yellow; juv greener with yellow cheeks. Characteristic 'tsee-tsee-tsee-tsit'; song starts similarly, ends in liquid trill. Woods, hedges, orchards, gardens. Nest of moss with felted cup, usually in hole in tree, wall, 7–16 eggs, Apr–Jul. Food insects, spiders, seeds, fruit, scraps. T, ex Ic, nFS. [**8**]

Great Tit *Parus major* PARIDAE L 13·5–14·5 cm. Glossy blue-black head with white cheeks, black extending as band down yellow underparts; juv duller, paler, with brown cap, yellowish cheeks. Huge vocabulary, louder, more metallic than other tits; song ringing 'teecha-teecha'. Woods, parks, timbered hedgerows, gardens. Nest of moss with felted cup, in any cavity, 5–12 eggs, Apr–Jul. Food insects, slugs, snails, earthworms, fruit, seeds. T, ex Ic. [ad **9**] [juv **10**]

Nuthatch *Sitta europaea* SITTIDAE L 13·5–14·5 cm. Bill long, tail short; upperparts blue-grey, underparts buff (WE, CE) to white (SC) with chestnut flanks; juv lacks chestnut. Climbs trunks, descends headfirst. Ringing 'chwit, chwit, chwit'; song loud piping or ringing notes, rapid trill. Woods, parks, gardens, with old trees. Nest of bark flakes, dead lvs, in hole in tree, wall, entrance reduced by plastered mud, 6–9 eggs, Apr–Jun. Food nuts, acorns, seeds, insects. T, ex nBr, Ir, Ic, Fi, nSC. [WE **11**]

Treecreeper *Certhia familiaris* CERTHIIDAE L 12·5–13 cm. Bill thin, curved, tail stiff; buff-streaked brown above, silvery-white below (paler and whiter eGe, Po, FS); juv rufous-tinged, more spotted. Climbs trunks with tail as prop. Woods, parks. Song thin: 2–4 notes, flourish at end. Cup nest of twigs, wood chips, roots, grass, moss, lined bark, feathers, hair, wool, behind loose bark, ivy, or in crack in trunk, 5–7 eggs, Apr–Jul. Food insects, larvae, spiders, woodlice. T, ex Ic, wFr, Ne, nFS, rare Be. [WE **12**]

Short-toed Treecreeper *Certhia brachydactyla* CERTHIIDAE L 12·5–13 cm. Very like treecreeper, but duller and greyer above, dirtier below, with less distinct superciliaries. Calls and song louder, less sibilant, more emphatic. Open woodland preferred. Nest, food similar. Fr, Lu, Be, Ne, seDe, Ge, Cz, Po. [**13**]

1
2
3
4
5
6
7
8
9
10
11
12
13

Golden Oriole *Oriolus oriolus* ORIOLIDAE L 23–25 cm. ♂ bright yellow with mainly black wings, tail; ♀ and juv yellow-green above, browner wings and tail, bill pink. Loud, fluty whistle 'weela-weeo'; harsh, jay-like notes. Open woods, parks, orchards, riverside trees. Nest hammock of grass, bark strips, wool, bound into horizontal fork, 3–5 eggs, May–Jul. Food insects, fruit, berries, also slugs, snails. Summer Fr, Lu, Be, Ne, De, Ge, Cz, Po, Fi, rare Br, sSw. [♂ **1**] [♀ **2**]

Jay *Garrulus glandarius* CORVIDAE L 33–35 cm. White rump contrasts with black tail in flight; mainly pink-brown (back greyer on Continent), with streaked crown, black moustache, white and black-barred blue wing-patches. Raucous 'skraaak'. Woods, young plantations, scrub, parks, orchards. Nest of twigs, lined roots, in fork, 5–7 eggs, Apr–Jul. Food acorns, peas, nuts, fruit, berries, nestling birds, eggs, slugs, snails, insects. T, ex Ic, nFS. [**3**]

Magpie *Pica pica* CORVIDAE L 44–47 cm. Boldly pied, glossed blue, purple, green, with long tail. Harsh, rattling chatter. Wood-edges, farmland with tall hedgerows, grassland with trees, orchards, scrub. Nest of sticks, lined mud, then roots, grass, usually with canopy of criss-crossed thorn twigs leaving side-entrance, in tree, tall hedge, thorn bush, locally in gorse, rarely on house or telegraph-pole (FS), 5–8 eggs, Mar–Jul. Food insects, small mammals, nestling birds, eggs, snails, slugs, other invertebrates, corn, nuts, fruit, berries. T, ex Ic. [**4**]

Jackdaw *Corvus monedula* CORVIDAE L 32–34 cm. Black above with grey to whitish nape, grey-black below; juv browner. 'Chak' (or 'jack'), sometimes in series, also 'kia'. Woods, parks, farmland or scrub with old trees, also old buildings, towns, quarries, sea-cliffs. Colonial; nest of sticks, lined wool, grass, in cavity in tree, building (*eg* chimney), rocks, also in thick ivy, bottom of heron nest, rabbit burrow, 4–6 eggs, Apr–Jun. Food insects, other invertebrates, cereals, berries, birds' eggs and nestlings. T, ex Ic, nFS. [**5**]

Rook *Corvus frugilegus* CORVIDAE L 44–47 cm. Glossy black, with bare, grey-white face; imm duller, with feathered face; distinguished from carrion crow by more slender, pointed bill, shaggy 'trousers'. Typically 'caw' or 'kaah'. Arable grassland with clumps of trees, copses, also urban parks, town edges. Colonial; nest of sticks, lined lvs, grass, moss, high in tree, 3–6 eggs, Mar–Jun. Food cereals, roots, potatoes, insect larvae, earthworms, snails, slugs, also young birds, eggs. Br, Ir, Fr, Lu, Be, Ne, De, Ge, Cz, Po, local sSC, only summer sFi. [ad **6**] [imm **7**]

Carrion/Hooded Crow *Corvus corone* CORVIDAE L 46–48 cm. Bill stout, slightly curved; carrion crow all-black glossed green; hooded crow has grey back and underparts, juv rather browner; intermediates where ranges overlap. Harsh, croaking 'kraah'; also motor-horn note. Wood-edges, farmland with timbered hedgerows, also urban squares, heaths, moors, sea-cliffs. Nest of sticks, lined wool, hair, in tree, on telegraph-pole, cliff-ledge, 4–6 eggs, Mar–Jul. Old nests used by many other birds. Food carrion, small mammals, birds, eggs, frogs, slugs, snails, insects, other invertebrates, also grain, seeds, acorns, fruit. Carrion Br, Fr, Lu, Be, Ne, sDe, wGe, wCz; Hooded nBr, Ir, De, Ge, Cz, Po, FS, to WE in winter. [Carrion **8**] [Hooded **9**]

Starling *Sturnus vulgaris* STURNIDAE L 21–22 cm. Plump, tail short, wings pointed; bill long, sharp, brown turning yellow in spring; all-blackish in summer, glossed green, purple; spangled with buff above and white below in winter, esp ♀, imm; juv mouse-brown with whitish throat. Harsh 'tcheerr'; rambling song of warbles, whistles, clicks, rattles, also mimicking other birds and noises. Woods, farmland, towns, moorland edges, sea-cliffs. Untidy nest of straw, grass, lvs, lined feathers, wool, moss, in hole in tree, wall, building, cliff, 4–7 eggs, Apr–Aug. Food insects, earthworms, other invertebrates, seeds, berries, fruit, cereals. T, but only summer eGe, Cz, Po, Fi, and all but swSC. [summer **10**] [winter **11**] [juv **12**]

Chaffinch *Fringilla coelebs* FRINGILLIDAE L 15–15·5 cm. White shoulders, wing-bar, tail-edges; ♂ slate-blue crown and nape, chestnut back, greenish rump, pinkish underparts; ♀ and juv yellow-brown, lighter below, juv less green on rump. Distinctive 'pink-pink'; song of quickening notes with terminal flourish 'cheweeoo'. Hedgerows, woods, orchards, gardens, scrub. Cup nest of moss with grass, fibres, adorned spiders' webs, lichens, bark flakes, lined feathers, wool, hair, in bush, hedgerow, creeper, tree, 4–5 eggs, Apr–Jul. Food seeds, corn, insects, earthworms. T, ex Ic. [♂ **1**] [♀ **2**]

Brambling *Fringilla montifringilla* FRINGILLIDAE L 14–15 cm. In lowlands only in winter, when distinguished from chaffinch by white rump, less white in wings, tail; ♂ then has mottled brown head and back (glossy black in summer), orange-buff shoulders and breast; ♀ and juv dull brown above, but chestnut-buff on breast. Hoarse 'tchuc'. Breeds in birch scrub, open conifer forest, but winters beech-woods, stubble, root-crops, stackyards. Winter food beechmast, corn, seeds, berries. Summer only nFS; winter T, ex Ic, Fi, all but sSC. [winter ♂ **3**]

Serin *Serinus serinus* FRINGILLIDAE L *c*11·5 cm. Tiny, bill stubby, tail short, forked; ♂ ♀ rump yellow, ♂ also yellow on head and breast, esp summer; juv browner, heavily streaked, rump brownish. Canary-like 'tsooeet', twittering flight-call; song rapid jangle. Cultivated areas, copses, scattered trees, wood-edges, gardens. Nest of grass, stalks, moss, roots, lichens, bound with spiders' webs, lined fine roots, plant down, feathers, hair, usually in outer fork of branch, 3–5 eggs, Apr–Jul. Food largely seeds. Fr, Lu, Ge, Cz, Po, scarce Be, Ne, De, has bred sBr, sSw. [♂ **4**] [♀ **5**]

Greenfinch *Carduelis chloris* FRINGILLIDAE L 14–15 cm. Plump, bill heavy, whitish, tail short, forked; ♂ olive-green with yellower-green rump, underparts, bold yellow patches on wings, tail; ♀ duller, greyer, with less yellow; juv browner, more streaked, still yellow on wings, tail. Rapid twitter, short 'chup', canary-like 'tsooeet'; song combines warbling twitter with nasal 'tsweee'. Wood-edges, hedgerows, parkland, gardens. Cup nest of moss, wool, grass, woven round twig base, lined roots, hair, sometimes feathers, usually in bush or small tree, 4–6 eggs, Apr–Aug. Food grain, seeds, berries, insects in summer. T, ex Ic, nFS. [♂ **6**]

Goldfinch *Carduelis carduelis* FRINGILLIDAE L *c*12 cm. Wings and tail black, white-spotted at ends, with broad yellow wing-bar; face red, contrasting with whitish bill, black and white head; head of juv grey-buff, streaked brown. Canary-like, liquid twitter. Orchards, parks, avenues, hedgerows, wood-edges, gardens; in winter, waste ground, rough pastures. Cup nest of roots, dead grass, wool, lined plant down, wool, usually in outer fork of tree-branch, 5–6 eggs, May–Aug. Food seeds, esp thistles, some insects. T, ex Ic, but only sFi, sSw, seNo. [ad **7**] [juv **8**]

Linnet *Carduelis cannabina* FRINGILLIDAE L 13–13·5 cm. ♂ head greyish, back chestnut, wings and tail black-brown edged white; in summer, forecrown and breast crimson; ♀ and juv duller, more streaked, no red. Call 'tsooeet', flight-note rapid twitter; song twittering combination of musical and twanging notes. Hedgerows, plantations, commons, heaths, coastal dunes. Social; cup nest of stalks, grass, moss, twigs, lined wool, hair, in gorse, bramble, thorn, young conifer, 4–6 eggs, Apr–Sep. Food seeds, some insects, esp in summer. T, ex Ic, nFS, but only summer Fi and all but swSC. [♂ **9**] [♀ **10**]

Bullfinch *Pyrrhula pyrrhula* FRINGILLIDAE L 14–15 cm. Stout black bill, black cap, white rump contrasting with black tail, black wings with white bars; ♂ blue-grey and pink, ♀ grey-brown and pink-brown; juv browner, no black cap. Low, piping 'peu'. Woods with undergrowth, thick hedgerows, young plantations, orchards. Cup nest of fine twigs, lined dark rootlets, often hair (hairy lichen in FS), in dense bush, evergreen tree, 4–6 eggs, Apr–Aug. Food buds, seeds, berries, insects in summer. T, ex Ic, nFS. [♂ **11**] [♀ **12**]

1
2
3
4
5
6
7
8
10
11
12
9

Hawfinch *Coccothraustes coccothraustes* FRINGILLIDAE L 16–17 cm. Huge bill (blue in spring, yellowish in winter), bull-neck, short tail with white tip; grey hindneck, black throat, whitish patches on blue-black wings; juv throat yellow, underparts spotted and barred. Clipped, explosive 'ptzik'. Open woodland, parks, orchards. Cup nest of twigs, bark, lined roots, grass, hair, moss, lichen, usually near end of horizontal branch, 3–6 eggs, Apr–Jul. Food kernels, large seeds, insects in summer. T, ex nBr, Ir, Ic, Fi, all but sSC. **[1]**

House Sparrow *Passer domesticus* PASSERIDAE L 14–15 cm. ♂ dark grey crown and nape, chestnut at sides, black bib, whitish cheeks; ♀ and juv duller, dark-streaked brown above, grey-brown below. Noisy: chirping, twittering calls. Farms, villages, towns, feeding on cultivation. Often colonial; untidy domed nest of straw, grass, lined feathers, wool, rubbish, in ivy, on building, high in hedgerow or tree, or cup nest (little more than lining) in cavity, 3–6 eggs, Apr–Aug. Food grain, seeds, buds, scraps, insects, earthworms. T, ex Ic. **[♂ 2] [♀ 3]**

Tree Sparrow *Passer montanus* PASSERIDAE L 13–14 cm. Smaller, neater than house sparrow, sexes alike; differs in chocolate crown, neat black bib, black spot on purer white cheeks; juv has duller crown, greyer-black bib and spot. Metallic 'chip', flight-call hard 'tek, tek'. Timbered hedgerows, open woods, parks, orchards, riversides, cliffs, towns. Often colonial; nest of straw, grass, moss, lined feathers, in hole in tree, thatch, building, quarry, or more flask-shaped in open, 4–6 eggs, Apr–Aug. Food seeds, corn, insects, spiders. T, ex Ic, most Fi, nSC. **[4]**

Yellowhammer *Emberiza citrinella* EMBERIZIDAE L 16–17 cm. ♂ mainly yellow head, underparts, chestnut rump; ♀ less yellow, with dark head-markings; juv darker, hardly yellow, less chestnut rump. Ringing 'twink', short 'twick', more liquid 'twitic'; song rapid series of tinkling notes, prolonged ending. Arable and grass with hedgerows, young plantations, roadsides,

bracken-clad hills, heaths. Cup nest of straw, stalks, moss, lined fine grass, hair, on or near ground in bank, hedge-base, grass-grown bush, brambles, 3–5 eggs, Apr–Sep. Food corn, seeds, fruit, insects, other invertebrates. T, ex Ic, but only summer FS. [♂ **5**] [♀ **6**]

Cirl Bunting *Emberiza cirlus* EMBERIZIDAE L 15·5–16 cm. ♂ distinguished from yellowhammer by face-pattern, dark crown, greenish breast-band with chestnut at sides; ♀ and juv only by olive-brown rump. Thin 'sip', flight-note 'sissi-sissi-sip'; song trilling rattle. Bushy hillsides, scrub with trees, hedgerows, parks, open fields in winter. Cup nest of roots, grass, moss, sometimes lvs, lined fine grass, hair, usually off ground in hedge, gorse, bramble, low tree, 3–4 eggs, May–Sep. Food corn, seeds, berries, insects. sBr, Fr, Lu, swGe, rare Be. [♂ **7**] [♀ **8**]

Ortolan *Emberiza hortulana* EMBERIZIDAE L 15·5–16 cm. ♂ grey-green head and breast, yellow eye-ring and throat, pink-buff underparts, reddish bill and legs; ♀ duller, with streaks on breast; juv brown, streaked black, but still eye-ring, reddish bill. Soft 'tsip', shrill 'tseeip', piping 'tseu'; song *c*6 musical notes, lower-pitched at beginning or end. Open country with scrub, scattered bushes, lowlands to mountains, also wood-edges, esp birch. Cup nest of grass, roots, lined fine roots, hair, on or near ground in grass, corn, rough vegetation, 4–6 eggs, May–Aug. Food seeds, insects, slugs, snails. Summer. T, ex Br, Ir, Ic, De, nSC. [♂ **9**] [♀ **10**]

Corn Bunting *Miliaria calandra* EMBERIZIDAE L 17–18 cm. Large, thickset, with big head, heavy bill; all brown, streaked black. Often flies with dangling legs. Harsh 'chip'; song high-pitched, accelerating, ending in trill, like jangling keys. Arable and grassland, often without hedgerows, also downs, gorse commons, coastal scrub. Cup nest of grass, lined roots, hair, in long grass or adjacent crop, in low brambles, marram, sometimes in bush, 3–5 eggs, May–Sep. Food grain, seeds, fruits, insects, other invertebrates. T, ex Ic, Fi, No, but only swSw. [**11**]

Mole *Talpa europaea* TALPIDAE BL 120–150 mm, TL 20–35 mm. Velvety near-black fur, broad spade-like front feet and absence of ear-pinnae diagnostic; eyes minute, hidden in fur; tail often carried erect. Pasture, arable fields, woodland. Subterranean, rarely seen above ground; mole-hills hemispherical, without opening or vegetation, unlike ant-hills. Feeds on earthworms, insects and other invertebrates. T, ex Ir, Ic, nFS. [1]

Hedgehog *Erinaceus europaeus* ERINACEIDAE BL 225–275 mm. Sharp spines on upperside; pointed snout; truncated head and neck; short legs. Grunts quietly; often detected by rustling among dead leaves. Hedgerows, woods, esp where damp grassland nearby; needs cover for nest. Feeds on insects, slugs, snails, also mice, frogs, berries. Nocturnal; by day after rain. Rolls into ball if frightened. Hibernates, usually from Oct to early Apr, in moss or leaf-lined hole among shrubs, occasionally in rabbit burrow or shed. T, ex Ic, nFS. [2]

Common Shrew *Sorex araneus* SORICIDAE BL 60–85 mm, TL 30–55 mm, HF 12–14 mm. Snout pointed, as in all shrews. Adults very dark brown on back, medium brown on flanks, greyish below, tail almost naked; juv lighter brown, tail well-haired, especially at tip; teeth red-tipped. (Smaller pygmy shrew *S. minutus* is colour of juv common shrew, with relatively longer tail; larger water shrew *Neomys fodiens* is black above, with keeled tail; both widespread.) Makes high-pitched squeaks. Woodland, hedgerows, long grass, on surface and in shallow tunnels. Feeds on insects, woodlice, small worms. T, ex Ir, Ic. [3]

Common White-toothed Shrew *Crocidura russula* SORICIDAE BL 65–95 mm, TL 35–45 mm, HF 11–13 mm. Greyish-brown, faintly speckled, grey below, without sharp demarcation on flank; tail with scattered long whiskers; teeth entirely white; ears more prominent than in common shrew. (Lesser white-toothed shrew *C. suaveolens*; Fr, Ge; is slightly

smaller (BL 50–80 mm, TL 25–40 mm, HF 10–12 mm) but very similar.) Habitat, food and voice as common shrew. Fr, Lu, Be, Ne, Ge, Cz, Po. [4]

Bicoloured Shrew *Crocidura leucodon*
SORICIDAE BL 65–85 mm, TL 30–40 mm, HF 12–13 mm. Very similar to common white-toothed shrew (teeth white, whiskers on tail), but darker above and paler below, with sharp demarcation on flank. Habitat, food and voice as for common shrew. Fr, Lu, Be, Ne, Ge, Cz, Po. [5]

Bank Vole *Clethrionomys glareolus*
CRICETIDAE BL 90–110 mm, TL 40–50 mm. Distinguished from other voles by rich chestnut upperparts (less distinct in juv), larger eyes and ears, longer, hairier tail. Dominant vole in woodland, also in hedgerows, bracken, scrub, deep heather. Diurnal, often seen climbing on logs or brushwood. Feeds on leaves (*eg* bramble), buds, fruit, seeds. Common prey of tawny owl. T, ex Ic, nFS. [6]

Common Vole *Microtus arvalis*
CRICETIDAE BL 90–120 mm, TL 30–45 mm. Eyes and ears smaller than in bank vole; tail about one third length of head and body, only faintly darker above than below. (Field vole *M. agrestis* very similar but fur shaggier, ears hairier and tail more clearly two-coloured; T, ex Ir, Ic. Water vole *Arvicola terrestris* similar but much larger (BL <220 mm), often found away from water; T, ex Ir, Ic.) Abundant in grassland, including quite short pasture where extensive tunnels made just below surface. Feeds on grass, rushes; droppings olive green. Preyed upon by short-eared owl, kestrel, harriers, buzzards and many other predators. T, ex Ir, Ic, FS, Br, but present Orkney and Guernsey. [7]

Pine Vole *Pitymys subterraneus*
CRICETIDAE BL 75–105 mm, TL 25–35 mm. Like common vole but eyes and ears smaller, colour rather greyer. Grassland, open woodland, making extensive tunnels near surface. Feeds on grass, roots. Fr, Lu, Be, sNe, sGe, Cz, sPo. [8]

Hazel Dormouse *Muscardinus avellanarius* MUSCARDINIDAE BL 60–90 mm, TL 55–75 mm, HF 15–18·5 mm. Orange-brown fur and small size distinguish from other dormice, and bushy tail from other mice. Woodland with dense undershrubs; summer nests in shrubs, makes use of bird nest-boxes; hibernates Oct–Mar in nest usually at ground level. Strictly nocturnal, active mainly in shrub layer; feeds on buds, fruit, nuts. sBr, Fr, Lu, Be, Ge, Cz, Po, sSw. [1]

Fat Dormouse *Glis glis* MUSCARDINIDAE BL 130–190 mm, TL 110–150 mm, HF 24–33 mm. Larger and greyer than other dormice, but only about half size of grey squirrel *Sciurus carolinensis* (Br, Ir). Woodland, orchards; nest in cavity of tree, sometimes in nest-boxes; hibernation nest often underground. Nocturnal; feeds on buds, fruit, nuts; enters buildings, esp for stored fruit. Fr, Lu, Ge, Po, Cz, (Br, Chilterns only). [2]

Garden Dormouse *Eliomys quercinus* MUSCARDINIDAE BL 100–170 mm, TL 90–120 mm. Medium-sized dormouse with tail tufted at end; black mask distinctive. Voice a churring noise. Woodland, hedgerows, scrub, gardens; nocturnal and arboreal but more often on ground than other dormice, nest often amongst rocks or in stone wall. Hibernates, enters buildings. Feeds on buds, fruit, insects. Fr, Lu, Be, Ge, Cz, Po. [3]

Wood Mouse *Apodemus sylvaticus* MURIDAE BL 80–110 mm, TL 70–115 mm, HF 20–25 mm. Yellowish-brown above, pale grey below with sometimes a small yellow mark on chest; hind feet long, slender and pale; eyes and ears large. Woodland, hedgerows, gardens, scrub; nests under logs, amongst tree-roots. Mainly on ground but climbs readily; feeds mainly on seeds, nuts, also buds, insects. Nocturnal. Preyed upon by owls, weasels, foxes. T, ex Fi, nSC. [4]

House Mouse *Mus musculus* MURIDAE BL 75–100 mm, TL 75–100 mm, HF 16–19 mm. Colour variable, from greyish-brown above and below to reddish-brown above and pale grey below when closely resembles wood mouse, but smaller, hind feet short, no yellow spot on chest, tail shorter. Fields, hedgerows, farmyards; also in buildings. Mainly nocturnal. Feeds largely on seeds but very versatile. T. [5]

Striped Field-mouse *Apodemus agrarius* MURIDAE BL 95–120 mm, TL 65–90 mm. Like wood mouse but narrow dark stripe along centre of back, no yellow spot on chest. (Birch mice *Sicista* spp. also striped but smaller, tail much longer than head and body; De, Ge, Cz, Po, FS.) Fields, edges of woods. Mainly nocturnal; feeds on seeds, buds, insects. De, Ge, Cz, Po. [6]

Yellow-necked Mouse *Apodemus flavicollis* MURIDAE BL 90–130 mm, TL 90–135 mm, HF 23–27 mm. Very similar to wood mouse but slightly larger, colour brighter above and paler grey below, yellow spot on chest larger, expanded sideways, sometimes forming complete collar. Habitat and behaviour as wood mouse but more confined to woodland. sBr, eFr, Lu, De, Ge, Cz, Po, sFS. [7]

Harvest Mouse *Micromys minutus* MURIDAE BL 60–75 mm, TL 50–70 mm, HF 13–16 mm. Smallest mouse; colour brighter and ears much shorter than in other long-tailed mice. Long grass, also in cereal crops; summer nest of grass woven into ball anchored to grass stems. Diurnal and nocturnal; feeds on seeds. T, ex Ir, Ic, SC. [8]

Common Hamster *Cricetus cricetus* CRICETIDAE BL 220–320 mm, TL 30–60 mm. Similar to familiar golden hamster but larger, black below, longer tail. (Golden hamster not wild in N Europe, but escapes may survive and breed.) Grassland and cultivation, living in burrows on edges of fields. Crepuscular and nocturnal; hibernates underground. Feeds on seeds which are stored in burrow. Lu, Be, Ge, Cz, Po. [9]

1
2
3
4
5
6
7
8
9

Lesser Horseshoe Bat *Rhinolophus hipposideros* RHINOLOPHIDAE BL 38–42 mm, TL 25–30 mm, FA 35–42 mm. Complex lobes of bare skin on muzzle, (shared only by greater horseshoe bat *R. ferrumequinum* which is much larger (FA 50–61 mm); sBr, Fr, Lu, Be, sGe, Cz; and Mediterranean horseshoe bat *R. euryale* (FA 43–49 mm); Fr only.) Lighter in colour than most bats. Mainly in wooded country; roost mostly in caves, sometimes in buildings, usually in small scattered colonies. All horseshoe bats roost by hanging freely, wings wrapped around body. Emerges after dark, flies with fluttering flight and very fast wingbeats. Feeds on flying insects (like all European bats). Hibernates Oct–Apr. sBr, Ir, Fr, Lu, Be, sNe, Ge, sPo, Cz. [1]

Whiskered Bat *Myotis mystacinus* VESPERTILIONIDAE BL 38–48 mm, TL 30–40 mm, FA 31–36 mm. Small, with dark greyish fur and all naked skin very dark, especially on muzzle. Tragus (lobe in conch of ear) narrow and pointed. (Brandt's bat *M. brandti* is very similar but adults are slightly more reddish above and yellowish below; T, ex Ir, Ic.) Mainly in or near woodland. Hibernates in caves and cellars, roosts in buildings and tree-holes in summer. Roosts in crevices or flat against wall, usually singly. T, ex Ic. [2]

Natterer's Bat *Myotis nattereri* VESPERTILIONIDAE BL 43–50 mm, TL 32–42 mm, FA 37–42 mm. Larger and paler than whiskered bat, tragus particularly long and slender. Fringe of hairs on margin of tail membrane, on either side of tail, is characteristic. Habitat and habits as whiskered bat. T, ex Ic, nFS. [3]

Greater Mouse-eared Bat *Myotis myotis* VESPERTILIONIDAE BL 68–80 mm, TL 50–60 mm, FA 57–67 mm. Largest bat in region. Narrow pointed tragus as in other *Myotis*. A colonial species, usually in caves in winter, caves and buildings in summer. Hangs freely from roof, usually crowded together. A local migrant. Fr, Lu, Be, Ne, Ge, Cz, Po, rare sBr. [4]

Noctule *Nyctalus noctula* VESPERTILIONIDAE BL 70–80 mm, TL 45–60 mm, FA 46–55 mm. Large size, golden-brown fur and narrow wings are diagnostic. Tragus is short and blunt. Squeaks loudly in roost and in flight. Wooded country, roosting in colonies in hollow trees; in winter also in buildings, rarely caves. Flies high, emerges early in evening. Migrates. T, ex Ic, Ir, nFS. [5]

Serotine *Eptesicus serotinus* VESPERTILIONIDAE BL 62–80 mm, TL 46–57 mm, FA 48–55 mm. Dull brown with broad wings. Similar to greater mouse-eared bat, but has smaller ears with short, blunt tragus. Open ground and fringes of woods. Roosts in buildings, hollow trees; in small colonies, hangs freely. Emerges soon after sunset; hesitant, fluttering flight. T, ex Ir, Ic, FS. [6]

Long-eared Bat *Plecotus auritus* VESPERTILIONIDAE BL 37–48 mm, FA 34–42 mm. Small; very long, conspicuous ears; broad wings. Voice sharp, shrill. Feeds on flying insects, also picks insects from foliage of trees. Largely nocturnal; spends day in hollow trees, lofts, steeples. Hibernates singly in caves, cellars, hanging from wall or roof. Some migrate long distances. T, ex Ic, nFS. [7]

Barbastelle *Barbastella barbastellus* VESPERTILIONIDAE BL 45–57 mm, TL 40–53 mm, FA 37–42 mm. Small and dark, with short ears that meet on top of head and very flattened wrinkled face. Wooded country but nowhere abundant. Roosts mainly in tree-holes, sometimes in buildings and caves, usually solitary. T, ex Ir, Ic, Fi, nSC. [8]

Pipistrelle *Pipistrellus pipistrellus* VESPERTILIONIDAE BL 33–45 mm, FA 28–35 mm. Smallest and commonest European bat; short, rounded ears; short, blunt tragus. Squeaks in flight. Emerges soon after sunset; fast, jerky flight. Gregarious, roosting in buildings or trees; hibernates in buildings, crevices in rocks. T, ex Ic, Fi, nSC. [9]

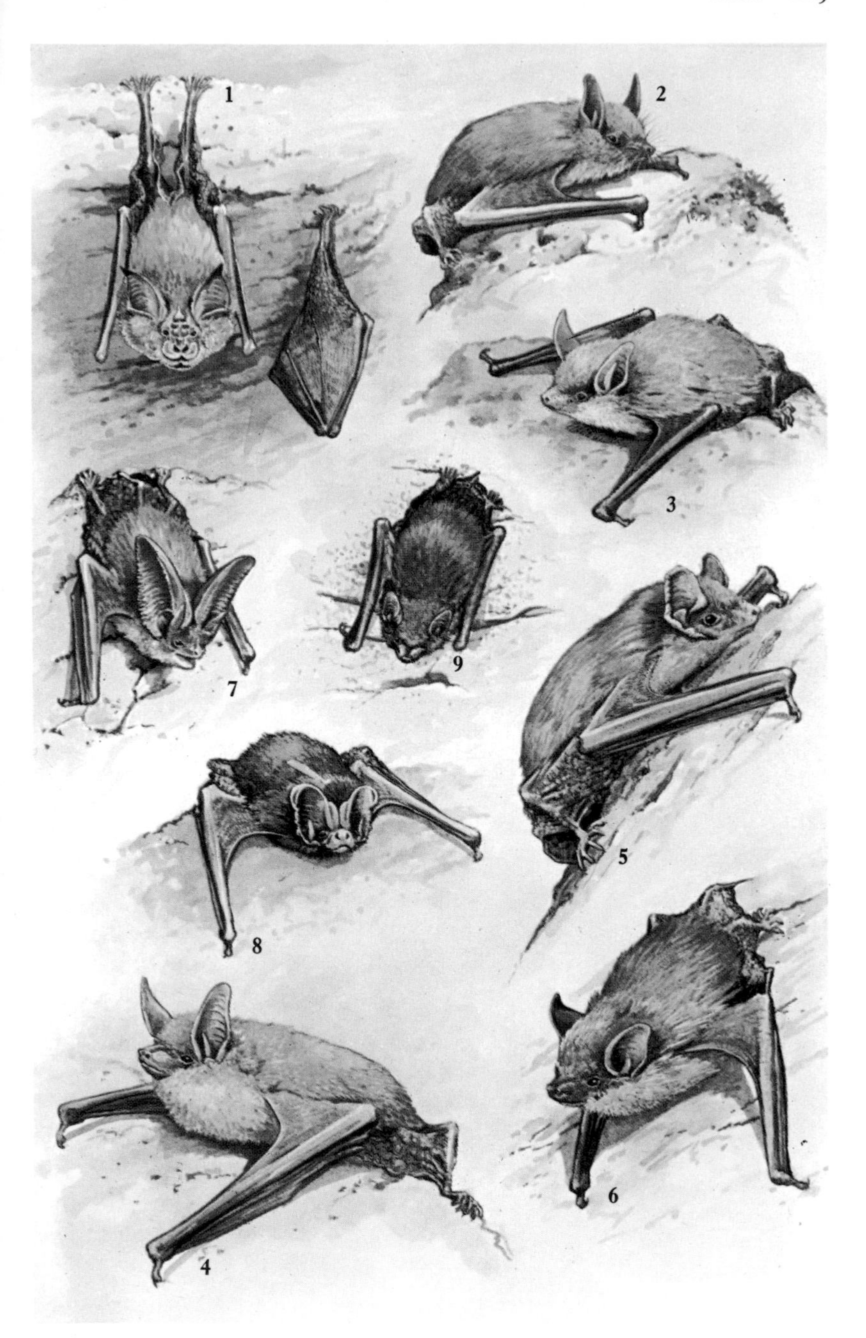
1
2
3
7
9
5
8
6
4

Polecat *Mustela putorius* MUSTELIDAE BL 35–45 cm, TL 13–18 cm. Fur dark above and below; sleek, showing paler underfur in places. (Pattern of pale marks on head distinguishes from similar mink *M. lutreola*; Fr, Ge, Fi; and American mink *M. vison*; T. Steppe polecat *M. eversmanni*; Cz, sPo; much paler, with dark feet. Feral ferrets also usually paler.) Woodland, hedges esp in marshes and river banks. Nocturnal. Feeds on rodents, frogs, birds. Fr, Lu, Be, Ne, De, Ge, Cz, Po, sFS, local wBr: Wales only. [1]

Stoat *Mustela erminea* MUSTELIDAE BL 20–30 cm, TL 8–12 cm, ♀ much smaller than ♂. Tail prominent, always with black tip; orange-brown above, white or yellowish below, demarcated by straight line; in N and E Europe turns pure white in winter, except tail-tip (ermine), in S and W whitening only partial. (Weasel *M. nivalis* is smaller, shorter tail lacks black tip; turns white only in far N and E; T, ex Ic.) Chatters; shrill alarm. All wood and open habitats with ground cover; nests in holes. Often hunts in groups, mainly at night; climbs well; feeds chiefly on rodents esp voles. Occasionally taken by large birds of prey. T, ex Ic. [2]

Beech Marten *Martes foina* MUSTELIDAE BL 40–47 cm, TL 22–26 cm. Larger than polecat with longer legs and tail, pointed ears; throat pure white, irregularly divided by dark marks. (Pine marten *M. martes* is similar but has the throat cream or yellow; T, ex Ic.) Makes shrill chattering call. Woodland, wooded farmland, rocky hillsides; an agile climber. Feeds on rodents and birds. Nocturnal. Nest in hollow tree or amongst rocks. T, ex Br, Ir, Ic, FS. [3]

Badger *Meles meles* MUSTELIDAE BL 65–80 cm, TL 13–18 cm. Black and white face-pattern unmistakable; body much less elongate than in other mustelids. Woodland, meadows, farmland. Burrows (sets) are extensive and very characteristic: usually a number of holes several

metres apart, each with a large mound of excavated material in front of it; sets usually in woodland or hedgerows. Feeds on wide variety of animal and vegetable food but esp earthworms. Nocturnal. T, ex Ic, nFS. [4]

Raccoon Dog *Nyctereutes procyonoides* CANIDAE BL 50–65 cm, TL 15–18 cm. Rather stocky, short-legged dog with greyish-brown fur, characteristic black pattern on face and tufts of long hair on cheeks. (Raccoon *Procyon lotor* PROCYONIDAE has similar facial pattern but larger ears and a ringed tail; (Lu, Be, Ne, Ge, from N America).) Woodland, scrub, riverbanks, marshes. Nocturnal, hibernates intermittently, den in burrow. Feeds on rodents, frogs, fish, carrion. (De, Ge, Cz, Po, FS, from E Asia). [5]

Red Fox *Vulpes vulpes* CANIDAE BL 55–75 cm, TL 30–50 cm. Upperside red-brown, underside whitish. Barking call, also high-pitched wailing. Woodland, farmland with hedges. Underground den, often enlarged rabbit hole. Eats rodents, rabbits, birds, frogs, earthworms, beetles. Nocturnal, usually solitary. T, ex Ic. [6]

Brown Hare *Lepus capensis* LEPORIDAE BL 50–65 cm, TL 7–11 cm, HF 11–15 cm. Long ears have black tips and tail has black stripe above. (More yellowish than rabbit; and mountain hare *L. timidus*, found on lowland farmland in Ir, which also lacks black on tail.) Farmland, open woodland; does not burrow. Some activity by day but mainly crepuscular. Eats grass, cereals, other crops; also tree bark in winter. T, ex Ic, nFS. [7]

Rabbit *Oryctolagus cuniculus* LEPORIDAE BL 35–45 cm, TL 4–8 cm, HF 7–10 cm. No black on ears, tail white below (raised when running). Ears and hind legs much shorter than brown hare. Farmland, wood-edges, dunes; colonial, extensive burrows, esp in sandy soil. Mainly crepuscular. Feeds on grass and herbs, including crops. Preyed upon by buzzards, foxes, cats. (T, ex Ic, Fi, nSC, from Spain.) [8]

Wild Boar *Sus scrofa* SUIDAE BL 120–160 cm, SH 80–100 cm. ♂ larger than ♀, with protruding, upturned tusks; young striped at first. Makes grunting and snorting noises. Deciduous woodland, sometimes emerging at night to feed in fields. Sexes separate most of year, ♂s usually solitary, ♀s usually with young, sometimes of two successive litters. Feeds on variety of vegetable and some animal material, in autumn and winter mainly on acorns, beechmast. Fr, Lu, Be, Ne, Ge, Cz, Po. **[1]**

Red Deer *Cervus elaphus* CERVIDAE BL <250 cm, TL 12–15 cm, SH <150 cm. Reddish-brown in summer, greyer in winter; fawn patch around tail, without black or white. Antlers in ♂s only, fully grown by August, shed in spring; fully adult ♂s have two points (tines) projecting forwards from near base of antler but young ♂s have only one. Calves spotted. Mainly broadleaved woodland; open mountain country in Br. Stags and hinds in separate herds most of year. Mature stags join herds of hinds and young in autumn when they roar and wallow in mud or peat. Feeds mainly by grazing, also browses foliage of trees and heather. T, ex Ic, Fi, nSC. [autumn ♂ **2**]

Fallow Deer *Cervus dama* CERVIDAE BL <150 cm, TL 16–19 cm, SH <110 cm. Colour and pattern very variable; mostly spotted in summer, unspotted and greyer in winter; uniformly dark or light brown forms are frequent. Long tail with black stripe distinctive, also black margin to white rump. Antlers only in ♂s, flattened at tips when well developed. Calves spotted. Deciduous woodland. Seasonal behaviour as in red deer. ♂s (bucks) make grunting rutting-call in autumn. Feeds by grazing and browsing, also acorns during winter. (T, ex Ic, Fi, nSC; from S Europe.) [autumn ♂ **3**]

Sika Deer *Cervus nippon* CERVIDAE BL <120 cm, TL 10–15 cm, SH <85 cm. Smaller than fallow deer, and lacks black stripe on tail; faintly spotted in summer, unspotted in winter. Antlers only in ♂s, similar to those of young red deer but lowest tine makes acute angle with main branch. Woodland, in small herds. Seasonal cycle as red deer. Rutting-call of ♂, in autumn, is sharp whistle. Grazes and browses. (Br, Ir, Fr, De, Ge; from E Asia.) [summer ♂ **4**]

Roe Deer *Capreolus capreolus* CERVIDAE BL 100–120 cm, SH 65–75 cm. Smaller than other native deer; pale rump conspicuous in winter, tail invisible; coat red-brown in summer, grey-brown in winter; fawns spotted at first. Only ♂ has antlers, rarely more than 3 points on each, shed in early winter, regrown by spring. Sharp bark. Woodland with good undergrowth, often emerges at night to feed on open ground. Solitary or in small groups. Browses broadleaved trees, shrubs; also eats grass, nuts, fruits. T, ex Ir, Ic. **[5]**

Muntjac *Muntiacus reevesi* CERVIDAE BL 80–100 cm, SH <50 cm. Very small deer; rounded back, unspotted (except calves). ♂s have very short simple antlers projecting backwards from permanent stalks (pedicels) and a pair of slender tusks protruding from upper jaw. (Chinese water deer *Hydropotes inermis* has even larger tusks but no antlers (sBr, Fr).) Voice a sharp bark, used by both sexes throughout year. Deciduous woodland. Browses on shrubs. (Br, Fr, from S China.) **[6]**

Mouflon *Ovis ammon* BOVIDAE BL 100–130 cm, TL 4–6 cm, SH 65–75 cm. Like domestic sheep but wool absent or concealed under normal hair, legs long; rump pale, tail dark. Pale patch on flank most prominent in mature ♂s. Horns small or absent in ♀s. Bleats. Open woodland. Most active at night, feeds mainly by grazing. (Fr, Lu, Be, Ne, Ge, Cz, from Corsica and Sardinia.) **[7]**

2
3
5
6
4
7
1

FURTHER READING

The following list gives at least one book per topic in the order of the ecological essay and the field guide.

The History of British Vegetation, W. Pennington (English Universities' Press, 1969, London)

Vegetation of the Earth in Relation to Climate and the Eco-physiological Conditions, H. Walter (English Universities' Press, 1973, London)

Warne's Natural History Atlas of Great Britain, A. Darlington (Frederick Warne, 1969, London)

Britain's Structure and Scenery, L. Dudley Stamp (Collins, 1946, London)

Blandford Colour Series: *eg Woodland Life*, G. Mandahl-Barth, A. Darlington (ed.) (Blandford Press, 1972, London)

Woodland Ecology, E. G. Neal (Heinemann Educational Books, 1953, London)

Hedges, E. Pollard, M. D. Hooper, and N. W. Moore (Collins, New Naturalist Series, 1974, London)

Grassland Ecology and Wildlife Management, E. Duffey, M. G. Morris, J. Sheail, L. K. Ward, D. A. Wells and T. C. E. Wells (Chapman and Hall, 1974, London)

The Biology of Weeds, T. A. Hill (E. Arnold, Studies in Biology Series, 1977, London)

Decomposition, C. F. Mason (E. Arnold, Studies in Biology Series, 1976, London)

Pests of Field Crops, F. G. W. Jones and M. G. Jones (E. Arnold, 1964, London)

Flora Europaea, 4 vols, T. G. Tutin (ed.) (Cambridge University Press, 1964–72, Cambridge)

The Wild Flowers of Britain and Northern Europe, R. Fitter, A. Fitter and M. Blamey (Collins, 1974, London)

The Concise British Flora, W. Keble Martin (Ebury Press and Michael Joseph, 1969, London)

A Field Guide to the Trees of Britain and Northern Europe, A. Mitchell (Collins, 1974, London)

Trees and Bushes of Europe, O. Polunin and B. Everard (Oxford University Press, 1976, London)

Flowers of Europe, O. Polunin (Oxford University Press, 1969, London)

Grasses, C. E. Hubbard (Penguin, 1968, London)

The Oxford Book of Invertebrates, D. Nichols, J. Cooke and D. Whiteley (Oxford University Press, 1976, London)

The Young Specialist looks at Molluscs, H. Janes (Burke, 1965, London)

Woodlice, S. L. Sutton (Ginn, 1972, London)

A Field Guide to the Insects of Britain and Northern Europe, M. Chinery (Collins, 1973, London)

Grasshoppers, Crickets and Cockroaches of the British Isles, D. R. Ragge (Warne, 1965, London)

A Field Guide to the Butterflies of Britain and Europe, L. G. Higgins and N. Riley (Collins, 1970, London)

The Moths of the British Isles, Two volumes, R. South (Warne, 1961, London)

Flies of the British Isles, C. W. Colyer and C. O. Hammond (Warne, 1968, London)

Land and Water Bugs of the British Isles, T. R. E. Southwood and D. Leston (Warne, 1959, London)

Beetles of the British Isles, Two volumes, E. F. Linssen (Frederick Warne, 1959, London)

Bees, Wasps, Ants and Allied Orders of the British Isles, E. Step (Frederick Warne, 1932, London)

Ray Society publications: *eg British Spiders*, G. H. Locket and A. F. Millidge (Ray Society, British Museum, 1951–53, London)

A Field Guide to the Reptiles and Amphibians of Britain and Europe, E. N. Arnold, J. A. Burton and D. W. Ovenden (Collins, 1978, London)

A Field Guide to the Birds of Britain and Europe, R. Peterson, G. Mountfort and P. A. D. Hollom (Collins, 1954, London)

Man and Birds, R. K. Murton (Collins, 1971, London)

The Atlas of Breeding Birds in Britain and Ireland, J. T. R. Sharrock (Poyser, 1976, Berkhamsted)

A Field Guide to the Mammals of Britain and Europe, F. H. van den Brink (Collins, 1967, London)

The Handbook of British Mammals, G. B. Corbet and H. N. Southern (eds.) (Blackwell, 1977, Oxford)

ACKNOWLEDGEMENTS

page	
6	Heather Angel
10	Heather Angel
14	M. Thonig/ZEFA
18	Adrian Warren/Ardea London
19 *left*	Heather Angel
right	A–Z Botanical Collection
22	Anthony and Elizabeth Bomford/ Ardea London
23	Derrick Boatman
26 *left*	Heather Angel
right	Jane Burton/Bruce Coleman Ltd
27	Heather Angel
33	A–Z Botanical Collection
37 *left*	Heather Angel
right	Denis Owen
40	Walter J. C. Murray/NHPA
44	E. A. Janes/Aquila
45	Derrick Boatman
49	Heather Angel
52	Jeremy Thomas/Biofotos
53 *left*	Neville Fox-Davies/ Bruce Coleman Ltd
right	Jane Burton/Bruce Coleman Ltd
56	David Hosking
61	A–Z Botanical Collection
65	Roger Hosking/NHPA

INDEX

Page references in **bold** refer to illustrations

1
storks
birds of prey

gamebirds
cuckoos

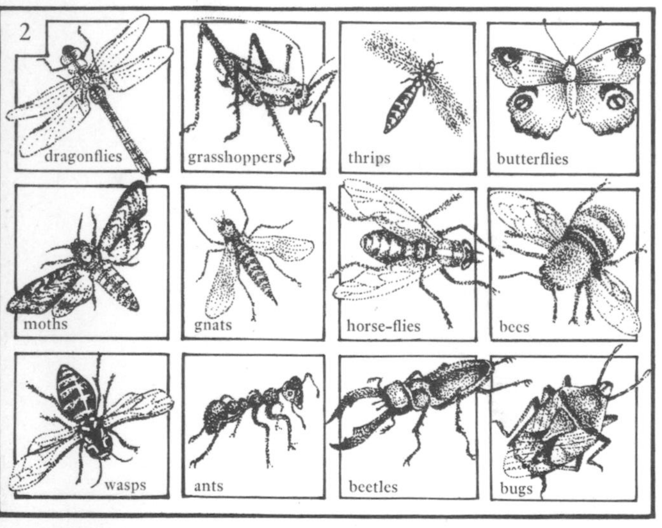
2
dragonflies
grasshoppers
thrips
butterflies
moths
gnats
horse-flies
bees
wasps
ants
beetles
bugs

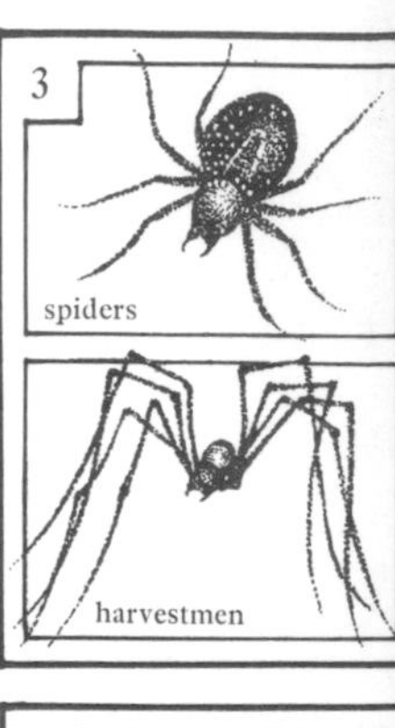
3
spiders
harvestmen

4

woodlice

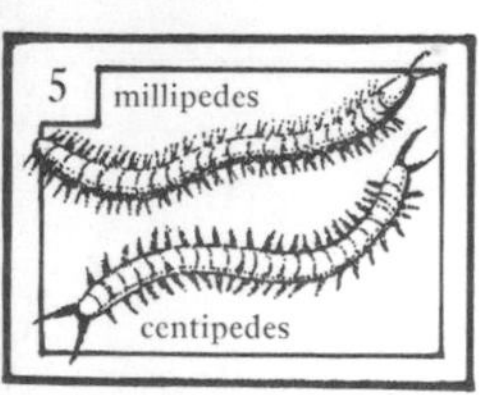
5
millipedes
centipedes

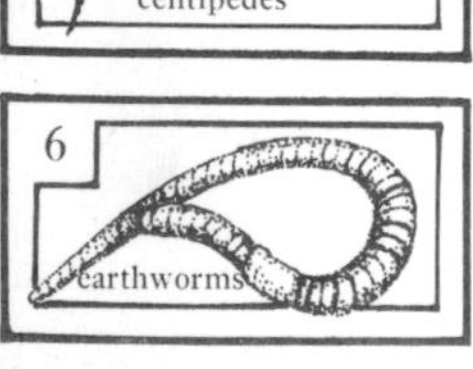
6
earthworms

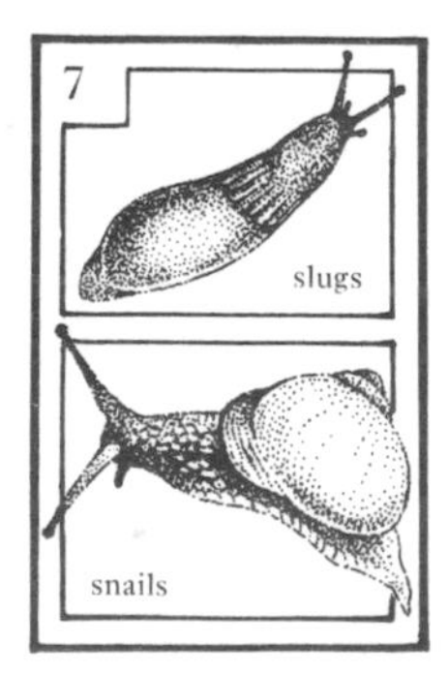
7
slugs
snails

8
newts

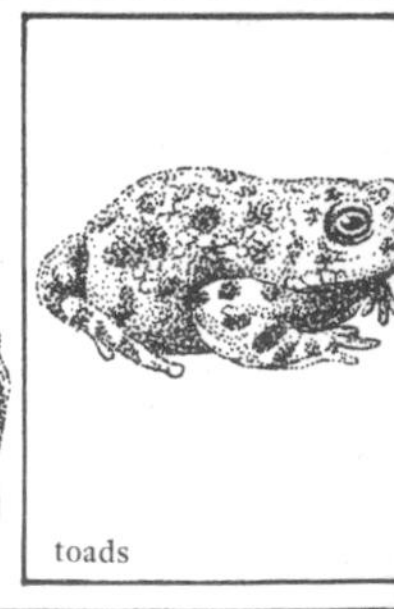
toads

9
snakes

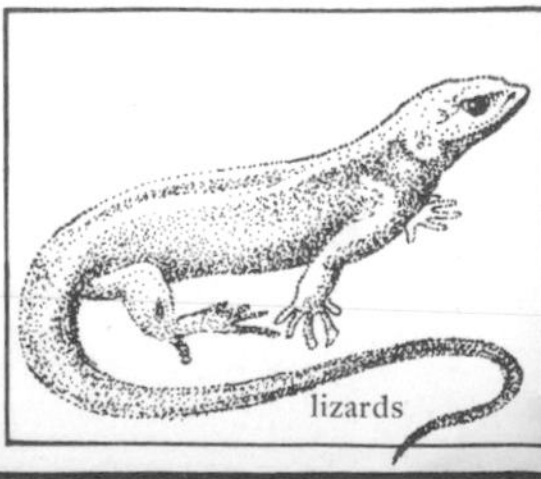
lizards